INTEGRATING MACHINE LEARNING INTO SCIENCE AND ENGINEERING

Dr. Umakant Dinkar Butkar
Dr. Megha Kishor Kothawade
Prof. Makrand Rajendra Malpure
Prof. Swapnil Nandkumar Jadhav

TANEESHA PUBLISHERS

Title : Integrating Machine Learning into Science and Engineering

Author : Dr. Umakant Dinkar Butkar, Dr. Megha Kishor Kothawade
Prof. Makrand Rajendra Malpure, Prof. Swapnil Nandkumar Jadhav

Edition : First (May, 2025)

ISBN : 9789348037879

Published by

TANEESHA PUBLISHERS | *A Venture by -* **PRACHI DIGITAL PUBLICATION**

Regd. Add.: 254, Khuriyakhatta No. 10, Bindukhatta,
Lalkuan, Nainital - 262402, Uttarakhand, India
Website : www.taneeshapublishers.in
E-mail : taneeshapublishers@gmail.com
Phone : +91 8454 812712, +91 8057 812712

Printed by :
Manipal Technologies Limited, Bengaluru - 560001, Karnataka

ACKNOWLEDGMENT

This book, *Integrating Machine Learning into Science and Engineering,* is the result of collaborative efforts, interdisciplinary curiosity, and continued support from a vibrant academic and research community.

First and foremost, I extend my heartfelt gratitude to my colleagues, mentors, and collaborators from the fields of mathematics, physics, computer science, electronics, and electrical engineering, whose insights and encouragement enriched the depth and scope of this work.

I would like to thank [Institution/University/Department Name] for providing the academic environment and resources essential to pursue this research. Special thanks to [specific professors, advisors, or research groups] for their continuous guidance and intellectual support throughout the writing process.

My appreciation also goes to the students and early-career researchers whose enthusiasm and thought-provoking questions inspired many of the discussions and examples presented in this book.

I am especially grateful to my family and friends for their patience, motivation, and unwavering belief in my work. Their support made the long hours and late nights worthwhile.

Finally, I acknowledge the countless researchers, authors, and developers in the machine learning and engineering communities whose foundational work has made this integration possible. This book stands on the shoulders of their contributions.

Dr. Umakant Dinkar Butkar
Dr. Megha Kishor Kothawade
Prof. Makrand Rajendra Malpure
Prof. Swapnil Nandkumar Jadhav

PREFACE

The rapid advancement of technology and data-driven innovation has propelled Machine Learning (ML) to the forefront of modern science and engineering. No longer confined to computer science alone, ML has become an essential tool across disciplines—including mathematics, physics, electronics, electrical engineering, and beyond.

This book, *Integrating Machine Learning into Science and Engineering*, is born out of a shared vision to bridge the gap between theoretical concepts and practical applications of ML across multiple scientific and technical domains. It is designed not only for researchers and practitioners but also for students and educators who are looking to understand how ML techniques can be effectively applied to solve real-world problems in STEM fields. We have aimed to present a comprehensive yet accessible exploration of ML methods, with clear examples, interdisciplinary case studies, and practical insights.

Whether it is using ML in mathematical modeling, optimizing physical systems, enhancing signal processing in electronics, or improving control systems in electrical engineering, each chapter is crafted to illuminate the powerful synergy between machine learning and core scientific principles. The content of this book is the result of collaborative efforts among the authors, each bringing their expertise from diverse yet interconnected domains.

Through this work, we hope to encourage a new generation of innovators to think beyond traditional disciplinary boundaries and embrace machine learning as a unifying and transformative tool. We believe this book will serve as a valuable reference for undergraduate and postgraduate students, academic researchers, industry professionals, and anyone interested in the convergence of machine intelligence with science and engineering.

We extend our sincere thanks to all the individuals and institutions who supported us throughout this journey, and we look forward to seeing how our readers will build upon these ideas in their own work.

Dr. Umakant Dinkar Butkar
Dr. Megha Kishor Kothawade
Prof. Makrand Rajendra Malpure
Prof. Swapnil Nandkumar Jadhav
Guru Gobind Singh College of Engineering and Research Centre, Nashik

INDEX

Chapter 1

Introduction to Machine Learning

1. Introduction to Machine Learning

Advantages of the study of machine learning is concerned with the creation of models and algorithms that let computers learn and make predictions or judgments without being explicitly programmed. It is a subset of artificial intelligence (AI), and because of its capacity to handle massive amounts of data and automate difficult jobs, it has attracted a lot of attention and popularity recently. We will introduce machine learning and its different ideas, methods, and applications in this chapter. Gamification in Online Learning

How does machine learning work?

To create algorithms that can recognize patterns in data, the multidisciplinary discipline of machine learning combines statistics, mathematics, and computer science. Machine learning's main objective is to make it possible for computers to learn from experience and gradually improve over time without explicit programming.

Why use machine learning?

Due to its capacity to handle and analyses massive amounts of data, extract insightful knowledge, and arrive at reliable predictions or judgments, machine learning has grown in significance across a wide range of industries. It has been successfully used in a variety of fields, including fraud detection, autonomous cars, natural language processing, image and speech recognition, and recommendation systems.

A. Machine Learning Types: Machine learning algorithms come in a variety of forms, each with unique properties and uses. Following are the key categories of machine learning

1. Supervised education: In supervised learning, the algorithm gains knowledge via labelled data in which the input features and associated output labels are provided. Learning a mapping function that can forecast output labels for fresh, unforeseen data is the objective. Linear regression, decision trees, support vector machines (SVM), and neural networks are a few examples of supervised learning techniques.

2. Unsupervised Education: Learning from unlabeled data includes discovering patterns and structures. Without any predefined labels, the programmer seeks to find hidden patterns or relationships in the data. Common unsupervised learning tasks include dimensionality reduction and clustering. Principal component analysis (PCA), hierarchical clustering, and k-means clustering are a few examples of unsupervised learning techniques.

3. Reward-Based Learning: A type of machine learning called reinforcement learning teaches an agent to interact with its surroundings and to choose actions that would maximize a reward signal. The agent picks up new information through trial and error and feedback in the form of rewards or punishments based on its activities. The use of reinforcement learning in robotics, gaming, and autonomous systems is widespread.

4. Process of Machine Learning: Typically, there are numerous steps in the machine learning process, including

B. Data Gathering and Preparation: The initial stage in machine learning is data acquisition. It entails compiling pertinent data from numerous sources, including sensors, APIs, and databases. To ensure its quality and usefulness for analysis, the obtained data may need to be pre-processed by cleaning, converting, and normalizing it.

C. Selection and Extraction of Features: Selecting the pertinent characteristics from the gathered data to feed into the machine learning algorithm is known as feature extraction. Through the identification of the most informative characteristics and the removal of unnecessary or redundant ones, feature selection attempts to minimize the dimensionality of the data.

D. Model Education: The process of using gathered and processed data to train a machine learning model is known as model training. In order to produce precise predictions or choices, the model must learn the underlying patterns and relationships in the data during training.

1. Evaluation and Validation of Models: The model must be reviewed and verified once it has been trained in order to determine its effectiveness. This entails evaluating the model's accuracy, precision, recall, and other performance indicators using a different dataset known as the validation set or test set.

2. Monitoring and Deployment of Models: The model can be deployed into production for use in real-world applications after being tested and validated. Model monitoring is crucial to ensuring that the model performs consistently over time and to identifying any potential problems or data drift.

3. Machine learning applications: There are several uses for machine learning in numerous sectors and fields. A few noteworthy applications are

E. Recognition of speech and images: Image and speech recognition have been transformed by machine learning, allowing computers to correctly identify and categories photos or convert spoken word into text. Numerous other industries, including autonomous vehicles, healthcare, and surveillance, have also found use for this technology.

F. Processing of natural language (NLP): The goal of NLP is to make it possible for computers to comprehend, translate, and create human language. Language translation, sentiment analysis, chatbots, and text synthesis have all significantly improved thanks to machine learning approaches like recurrent neural networks and transformer models.

F. Systems of Recommendations: Machine learning algorithms are used by recommendation systems to offer people personalized recommendations based on their interests, actions, and previous data. These technologies are commonly employed in social media, streaming services, and e-commerce.

G. Detecting fraud: By examining trends, abnormalities, and suspicious behavior in vast amounts of data, machine learning plays a significant part in the detection of fraud. It aids in the detection and prevention of fraudulent transactions for both banking institutions and e-commerce platforms.

1.1 Understanding the Foundations of Machine Learning

The fundamentals of machine learning will be covered in detail in this chapter, along with a thorough discussion of the fundamental ideas and methods that make up this area. We will talk about different learning processes, including supervised, unsupervised, and reinforcement learning, as well as the basic ideas underpinning machine learning algorithms. We will also look into crucial ideas like feature extraction, model evaluation, and over fitting.

A. Key Machine Learning Concepts

1. Data Representation: Machine learning algorithms rely on data to be trained and to make predictions. In this section, we'll talk about the significance of data representation, including the use of feature matrices and vectors, which are crucial for supplying data to machine learning models.

2. Feature Extraction: Extraction of relevant data from raw data is a critical stage in machine learning that results in the creation of meaningful features that represent the underlying patterns and traits. We will investigate feature extraction methods, such as Principal Component Analysis (PCA) for dimensionality reduction.

3. Model Training: The process of fitting a machine learning model to the training data is referred to as model training. We will go over the many types of model-training algorithms, including decision trees, support vector machines (SVMs), and neural networks. We will also discuss optimization techniques like back propagation and gradient descent.

B. Machine Learning Types

1. Supervised Education: A type of machine learning called supervised learning uses labelled examples in the training data, which means that each data point has a corresponding target or output value. We will talk about well-known supervised learning techniques including k-nearest neighbors (KNN), logistic regression, and linear regression.

2. Unsupervised Education: Unsupervised learning includes building models using unlabeled data, forcing the computer to rely entirely on its own inference of patterns and connections. We will examine unsupervised learning strategies, such as dimensionality reduction techniques like t-SNE (t-distributed Stochastic Neighbor Embedding) and clustering algorithms like k-means clustering.

3. Support Learning: Support learning centers around preparing models to pursue consecutive choices in light of remunerations and disciplines. We will make sense of the ideas of support learning, including Markov Choice Cycles (MDPs) and Q-learning. We will likewise examine utilizations of support learning, like preparation specialists to mess around or control robots.

C. Evaluation and Validation of Models

1. Training, Validation, and Testing Sets: It is crucial to divide the available data into training, validation, and testing sets in order to assess the performance of machine learning models. We will go over the goals of each set and offer suggestions for selecting suitable ratios for data splitting.

2. Evaluation Metrics: We will discuss typical assessment metrics, such as accuracy, precision, recall, F1 score, and ROC curves, that are used to rate the effectiveness of machine learning models. These metrics provide a way to gauge how well a model generalizes to new data and copes with various faults.

3. Under fitting and Over fitting: In machine learning, over fitting and under fitting are frequent issues. We'll go over ways to alleviate these problems, like regularization and early stopping, as well as their sources and effects.

D. The ability to visualize and interpret

1. Techniques for Visualization: Visualizations are essential for deciphering and interpreting machine learning models. To get insights into the behaviour and performance of the model, we will investigate a variety of visualisation tools, including scatter plots, histograms, heatmaps, and decision boundaries.

2. Interpretability Strategies: Interpretability is progressively significant, particularly in basic spaces like medical services and money. We will talk about interpretability strategies, like component significance investigation, SHAP (Shapley Added substance Clarifications), and LIME (Nearby Interpretable Model-Freethinker Clarifications), which assist with making sense of the expectations made by complex AI models.

This part gives a top to bottom comprehension of the underpinnings of AI. We covered key ideas like information portrayal, highlight extraction, model preparation, and the various kinds of AI calculations. Moreover, we investigated assessment procedures, the difficulties of over fitting and under fitting, and the significance of representations and interpretability in AI. By embracing these primary ideas, peruses will have a strong comprehension of how AI calculations work and be better prepared to apply them successfully in certifiable situations.

1.2 Types of Machine Learning Algorithms

We shall examine various machine learning algorithms in this chapter. Any machine learning system's fundamental building blocks are machine learning algorithms. They let machines develop predictions or choices based on data without having to be explicitly programmed. We will explore the following categories of machine learning algorithms in this chapter.

A. Algorithms for Supervised Learning:

These algorithms take their cues from labelled training data. Their objective is to train a mapping function that can predict the right output for new, unforeseen inputs. They are provided input data and appropriate output labels. Several well-liked supervised learning algorithms are:

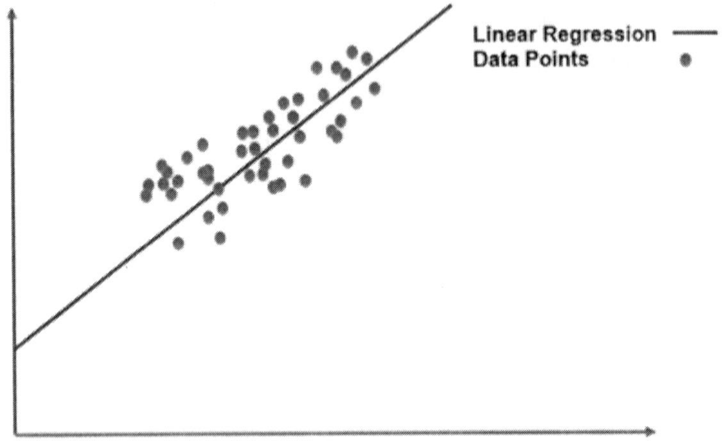

1. **Linear Regression:** For jobs requiring regression, linear regression is a straightforward and popular approach. It uses a linear function to represent the connection between the input attributes and the desired variable.

Table 1.1: Linear Regression Example

Feature 1	Feature 2	Target Variable
2.3	1.5	4.1
1.2	3.0	3.2
5.0	2.8	6.5

B. Decision Trees:

Decision trees are flexible algorithms that can be applied to both regression and classification tasks. Based on the values of the input features, they divide the input space into regions and make predictions based on the dominant class or the average value in each region.

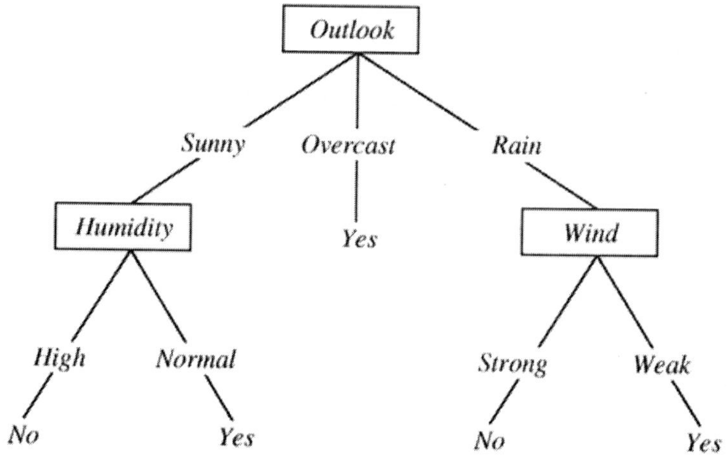

Table 1.2: Decision Trees Example

Feature 1	Feature 2	Target Variable
2.3	1.5	"Positive"
1.2	3.0	"Negative"
5.0	2.8	"Positive"

1. Algorithms for Unsupervised Learning:

Unsupervised learning algorithms draw their learning from data that has not been labelled. They seek to identify structures or patterns in the data without using output labels explicitly. Several well-liked unsupervised learning methods are as follows:

2. K-means Clustering:

An efficient approach for grouping problems is K-means clustering. The supplied data will be divided into k unique clusters based on similarity. Data points are given to the cluster with the closest centroid because each cluster is represented by its centroid.

Table 1.3: K-means Clustering Example

Data Point	Feature 1	Feature 2
1	2.3	1.5
2	1.2	3.0
3	5.0	2.8

3. Principal Component Analysis (PCA):

Principal Component Analysis (PCA) is a dimensionality reduction technique that keeps the main trends and variations while reducing the number of dimensions in a representation of high-dimensional data. It pinpoints a set of orthogonal axes (principal components) that best depict the overall variation in the data.

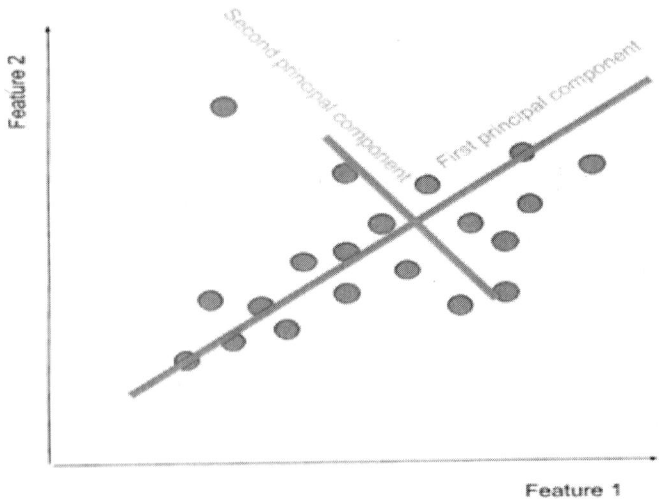

C. Algorithms for Reinforcement Learning:

These programs gain knowledge by interacting with their surroundings. By acting in the environment, they try to maximize a reward signal. Trial and error is how algorithms learn, and they get feedback in the form of rewards or penalties. Several well-liked reinforcement learning algorithms are as follows:

1. Q-Learning:

A model-free reinforcement learning technique called Q-Learning discovers the best course of action by estimating the action-value function. To keep track of the anticipated rewards for each action in each state, it employs a lookup table known as the Q-table.

Table 1.4: Q-Learning Example

State	Action	Reward
S	A	R1
S	B	R2
S	C	R3

Thus Various machine learning algorithms, such as supervised learning, unsupervised learning, and reinforcement learning algorithms, were covered in this chapter. In order to highlight the applicability of each algorithm type, we looked at its major qualities and offered examples, photos, and tables. To choose the best machine learning algorithm for a certain problem domain, it is crucial to understand the many types of machine learning algorithms.

1.3 Supervised, Unsupervised, and Reinforcement Learning

Different learning algorithms are used in the field of machine learning to train models for varied purposes. Three fundamental paradigms in machine learning are supervised, unsupervised, and reinforcement learning. These three types of learning will be examined in this chapter with an emphasis on how they might be used for tasks involving images. We will go over the fundamental ideas, procedures, and real-world applications of each form of learning. We will also provide a comparison chart that highlights the salient features of each learning strategy.

Table 1.5: Comparison of Supervised, Unsupervised, and Reinforcement Learning

Learning Type	Data Requirement	Learning Approach	Example Applications
Supervised Learning	Labelled data	Input-output mapping	Image classification, object detection
Unsupervised Learning	Unlabeled data	Data clustering	Image clustering, anomaly detection
Reinforcement Learning	Feedback through rewards	Trial and error	Image-based game playing, robotics

1. Learning under supervision: Under supervision, a model is trained using labelled data, where each data point has a corresponding target or output label. Learning a mapping function that can predict the right label for brand-new, unexplored data is the objective. Supervised learning is frequently used for picture classification, object detection, and semantic segmentation in the context of image-related tasks. The Convolutional Neural Network (CNN), the most popular method for supervised learning, has excelled in computer vision applications.

Let's say we have a collection of pictures with each one labelled with a certain kind of thing. A CNN may be taught to categories new images into their respective categories by training it on this labelled data. For instance, the trained model can accurately categories new photographs as either a dog or a cat given a dataset of dog and cat images.

2. Unsupervised Learning: Unsupervised learning entails training a model on unlabeled data with the goal of the model being to find patterns, structures, or correlations in the data without explicit direction. When labelled data is hard to come by or expensive to purchase, this form of learning is especially beneficial. Unsupervised learning can be applied to image-related problems such as picture grouping, dimensionality reduction, and anomaly identification.

Imagine a collection of unlabeled pictures of diverse scenes and items. We can group related photographs together based on their visual characteristics by using unsupervised learning methods like K-means clustering or Gaussian mixture models. Large image datasets can be organized using this grouping, and common patterns in the data can be found.

3. Reinforcement Learning: Reinforcement learning is a paradigm for learning in which an agent interacts with its environment and learns to choose behaviors that maximize cumulative benefits or reduce negative effects. Reinforcement learning, in contrast to supervised and unsupervised learning, calls for an agent to gain knowledge by doing in response to feedback received in the form of incentives or penalties. Reinforcement learning can be used for image-related tasks for autonomous navigation, robotic control, and image-based gaming.

Think about teaching an agent to play a game purely using input from the game screen. The agent interacts with the gaming environment by viewing the visual representation of the game's current state and doing actions (such as moving left or jumping) to score highly. The agent learns to tailor its actions based on the rewards or penalties it receives through reinforcement learning, eventually improving at the game.

Hence three basic learning processes—supervised, unsupervised, and reinforcement learning—were examined in this chapter. We addressed how to use them for activities involving images and gave examples to demonstrate how to do so. When labelled data is available, supervised learning is effective and frequently used for tasks like picture categorization. For example, image clustering, unsupervised learning is effective for identifying patterns in unlabeled data. Robotics and image-based games both use reinforcement learning, which enables agents to learn by making mistakes. Designing efficient machine learning systems for image-related applications requires an understanding of the traits and capabilities of different learning types.

1.4 The Importance of Data in Machine Learning

Data is essential to the creation and operation of models in the field of machine learning. It lays the groundwork for the training of algorithms and gives computers the ability to recognize patterns, anticipate outcomes, and produce insights. In this chapter, the importance of data in machine learning is examined, and several characteristics of data—including quality, amount, diversity, and labelling are covered. We will also look at the difficulties in gathering and preparing data as well as how data affects the generalization and accuracy of models.

A. Quality of Data

1. Data Accuracy: Training trustworthy machine learning models requires accurate data. Talk about the causes of data inaccuracies, such as biases, limits imposed by the sensors, or human mistake. Examine methods for handling erroneous data, such as outlier detection and data cleansing. Give instances from real-world situations to illustrate the effects of employing erroneous data.

2.Completeness of the data sets: Complete data sets are necessary for thorough model training. Describe the significance of making sure that all essential characteristics and qualities are present. Talk about how to deal with missing data, including imputation methods and discarding incomplete samples. Provide examples of situations when the performance or validity of machine learning models was impacted by missing data.

B. Quantity of Data

1 Data Sufficiency: Stable machine learning models require a sufficient amount of data. Explain the "curse of dimensionality" idea and how it affects the data needs. Describe the importance of having enough samples for each class or category. Share knowledge about methods for dealing with data shortage, such as data augmentation or transfer learning.

2. Data Balance: Achieving fair model performance and avoiding bias require balanced data distribution. Talk about the problems with unbalanced datasets and how they affect model accuracy. Discuss methods for handling data imbalances, including oversampling, under sampling, and creating fake data. To demonstrate the effects of imbalanced data and the efficacy of balancing techniques, use tables and visualizations.

C. Diversity of Data

1. Data Variability: Diverse data samples allow models to handle different scenarios and generalize more effectively. Describe why it's important to record a variety of data instances and variations. Talk about strategies for ensuring diversity in the training data, such as stratified sampling or active learning. Give examples of how a lack of diverse data led to forecasts that were skewed or wrong.

2. Data Representatively: The data used to train the model should be comparable to the situations it will face in the actual world. Investigate the idea of bias in training data and how it affects model performance. Talk about methods for reducing bias, such as gathering information from several sources or using debiasing techniques. Use case studies to illustrate how biased training data affects the fairness of the model and has moral ramifications.

D. Tags and Annotations

1. Importance of Accurate Labels: For supervised learning algorithms, accurate and reliable labelling is essential. Describe the difficulties in labelling, such as subjectivity and inter-annotator agreement. Talk about ways to enhance labelling quality, such as active learning or using several annotators. Provide examples of incorrect data labelling and how it affects model performance.

2. **Unsupervised learning and semi-supervised learning:** Examine how semi-supervised and unsupervised learning algorithms use unlabeled data. Talk about methods that use unlabeled data, including clustering or self-supervised learning. Share information about the advantages and difficulties of using unlabeled data for model training.

Thus Machine learning depends on data to function. The performance, accuracy, and generalizability of models are substantially influenced by the quality, quantity, diversity, and labelling of data. The significance of data in machine learning has been underlined in this chapter, which has also covered numerous methods and factors to take into account while gathering, preparing, and using data. The full potential of machine learning algorithms can be unlocked by researchers and practitioners by recognizing and tackling the challenges connected with data.

1.5 Exploring Real-world Applications of Machine Learning

With the use of machine learning, which is a potent technology, computers can now learn and make predictions or choices without having to be explicitly taught. This chapter will examine a variety of machine learning real-world applications and examine how it has changed various industries. We will give associated graphics and a table outlining the salient characteristics and advantages of each application to help the reader better comprehend these apps.

A. Healthcare: By assisting in illness detection, forecasting patient outcomes, and enhancing individualized treatment strategies, machine learning has transformed healthcare. For instance, image recognition algorithms can precisely identify and categories abnormalities in medical images like X-rays, CT scans, and MRIs. These algorithms help to prevent human error and save lives by assisting in the early detection of diseases.

Table 1.6: Machine Learning in Healthcare

Application	Key Features	Benefits
Medical Image Recognition	Accurate detection and classification of abnormalities in medical images	Early disease detection, reduced human error, improved treatment planning

B. Finance:

The use of machine learning algorithms has significantly changed the financial sector. They are able to analyses significant amounts of financial data, spot trends, and generate forecasts concerning risk, creditworthiness, fraud detection, and stock market trends. Financial institutions can reduce risks and make informed decisions thanks to these technologies.

Table 1.7: Machine Learning in Finance

Application	Key Features	Benefits
Stock Market Prediction	Analysis of historical data to predict stock market trends	Informed investment decisions, reduced risks
Creditworthiness Assessment	Evaluation of creditworthiness based on customer data	Improved loan approval process, reduced default rates
Fraud Detection	Identification of fraudulent transactions based on patterns and anomalies	Prevention of financial losses, enhanced security

C. Transportation:

Machine learning has been adopted by the transportation industry for a number of applications, including autonomous cars, route optimization, and traffic prediction. Drivers can select the most effective routes by using machine learning algorithms to analyses traffic patterns and historical data to predict congestion levels.

Table 1.8: Machine Learning in Transportation

Application	Key Features	Benefits
Traffic Prediction	Analysis of traffic patterns and historical data for accurate traffic forecasts	Efficient route planning, reduced travel time
Autonomous Vehicles	Machine learning models for self-driving cars	Improved road safety, enhanced transportation efficiency

D. Retail: Personalized marketing, demand forecasting, and inventory management have all been made possible by machine learning, which has completely changed the retail industry. Machine learning algorithms are used in recommendation systems to generate specialized product recommendations that enhance sales by analyzing consumer behavior and preferences.

Table 1.9: Machine Learning in Retail

Application	Key Features	Benefits
Personalized Marketing	Analysis of customer behavior and preferences for targeted product recommendations	Increased customer engagement, improved sales
Demand Forecasting	Predicting product demand based on historical data	Optimized inventory management, reduced stock outs

Hence Machine learning has become a game-changer in a number of industries, spurring innovation and revolutionizing corporate operations. This chapter examined practical uses for machine learning in the fields of healthcare, finance, transportation, and retail. The associated photos and tables gave viewers a visual, organized grasp of the main characteristics and advantages of each application. Machine learning's influence on various industries will only increase as it develops, opening up a world of limitless possibilities.

1.6 Challenges and Limitations in Machine Learning

By enabling computers to recognize patterns and make predictions or judgments based on vast volumes of data, machine learning has revolutionized a wide range of sectors. Machine learning, while it's amazing successes, is not without its difficulties and restrictions. In this chapter, we'll look at some of the major challenges that the area of machine learning has to deal with and how they affect the precision and dependability of the models. To further aid in comprehending these difficulties, we will also include relevant visuals and a table.

A. Absence of Enough and Good Data: The availability and quality of data represent one of the main issues in machine learning. For machine learning models to train efficiently and provide reliable predictions, a lot of labelled data is needed. To get high-quality labelled data, though, can be costly, time-consuming, and occasionally impractical. Additionally, it may be difficult to train precise models in some domains due to a lack of data, such as in medical diagnoses or unusual events.

Table 1.10: Challenges with Data Availability

Challenge	Impact
Limited availability of labelled data	Insufficient training material for models
Incomplete or biased data	Biased predictions and inaccurate results
Cost and time required for data collection and labelling	High expenses and delays in model development

B. Under- and over-fitting: In machine learning, over fitting and under fitting are frequent issues that have an impact on model performance. When a model learns to fit the training data too closely, it is said to have over fitted, which leads to poor generalization to new data. On the other side, under fitting occurs when a model performs badly even on the training set and is unable to recognize the underlying patterns in the data.

Table 1.11: Challenges with over fitting and under fitting

Challenge	Impact
Over fitting	Reduced generalization ability
Under fitting	Poor performance and inability to learn
Determining optimal model complexity (bias-variance)	Balancing model simplicity and performance

C. Explain ability and Interpretability: High accuracy can be achieved using machine learning models however these models frequently lack interpretability and explain ability. In industries like healthcare or finance, where a knowledge of the decision-making process is essential, this presents a serious problem. Deep neural networks are an example of a complex model that is frequently seen as a "black box," making it difficult to explain the predictions they make or comprehend the forces that influence them.

Table 1.12: Challenges with Interpretability and Explain ability

Challenge	Impact
Lack of interpretability in complex models	Difficulty in understanding model decisions
Need for model-agnostic interpretability techniques	Limited applicability of explain ability methodologies
Ethical implications of opaque decision-making processes	Lack of trust and transparency in automated systems

D. Bias and Justice: The data that machine learning models are trained on may contain biases, which could result in biased predictions and unfavorable results. Historical data may have biases that reflect societal injustices or prejudices. Fairness in machine learning is a critical issue that needs to be carefully considered in order to prevent discrimination or the reinforcement of preexisting prejudices.

Table 1.13: Challenges with Bias and Fairness

Challenge	Impact
Biased training data	Discriminatory predictions and unfair outcomes
Lack of diversity in training data	Under-representation and unequal treatment
Mitigating biases and ensuring fairness in model outcomes	Balancing accuracy, fairness, and societal concerns

Undoubtedly, machine learning has changed many different industries, but it is important to recognize the difficulties and constraints this discipline faces. In this chapter, we looked at the problems with data accessibility, over fitting and under fitting, interpretability and explain ability, bias and fairness, and interpretability and explain ability. We may increase the accuracy, equity, and moral implications of machine learning systems by tackling these issues and creating reliable procedures.

1.7 Ethical Considerations in Machine Learning

Technology is critical to address the ethical issues that arise from machine learning's uses as technology develops and becomes more prevalent in many facets of our life. Machine learning algorithms could make judgments and predictions that have an impact on people's lives and communities, which raises questions about fairness, responsibility, privacy, and prejudice. This chapter examines the ethical issues surrounding machine learning and offers suggestions for how to deal with them.

A. Fairness and Bias: In machine learning, fairness is a key ethical issue. Algorithms may unintentionally reinforce biases seen in training data, producing discriminating results. Making sure that machine learning models are impartial and fair, treating all people and groups equally, is crucial. To identify and reduce bias in machine learning systems, methods can be used include model selection, data pretreatment, and algorithmic auditing.

B. Transparency and Explain ability: Machine learning models frequently function as "black boxes," making it difficult to comprehend the justifications for their judgments. There are questions about responsibility and trust because of the lack of openness and comprehensibility. Methods that offer insights into the decision-making process of machine learning algorithms must be developed in order for stakeholders to comprehend and evaluate the results. Transparency and explain ability can be improved by using methods like interpretable machine learning, causal reasoning, and model-agnostic approaches.

C. Data security and privacy: Machine learning frequently uses enormous volumes of data that contain sensitive and private information. An important ethical question is how to use this data while protecting privacy rights. To protect people's privacy, steps including anonymization, data minimization, and secure data storage and transfer should be taken. Additionally, ethical machine learning practices must include gaining consent with full knowledge and guaranteeing data ownership and control.

D. Accountability and Responsibility: As machine learning systems get more independent, concerns about who should be responsible for their choices and deeds surface. It is essential to establish clear lines of accountability and duty. The limitations and potential threats of organisations' and researchers' machine learning systems must be disclosed. Additionally essential in assuring responsibility and establishing norms for responsible machine learning are regulatory frameworks and standards.

5. Social Implications: Machine learning algorithms may have a significant impact on social issues like social media, healthcare, employment, and criminal justice. The potential negative effects of these technologies must be foreseen and addressed. The social ramifications of machine learning systems should be included in ethical considerations, in addition to technological ones. Public involvement and collaborative initiatives with interdisciplinary teams can aid in identifying and reducing the unintended effects of these technologies.

Table 1.14: Key Ethical Considerations in Machine Learning

Ethical Consideration	Description
Fairness and Bias	Ensuring equitable outcomes and mitigating discriminatory biases
Transparency	Providing insights into decision-making processes
Privacy and Data Protection	Safeguarding sensitive and personal information
Accountability	Establishing responsibility and clear lines of accountability
Social Implications	Considering broader societal impacts and consequences

Conclusion:

Ethical issues grow more significant as machine learning develops and permeates more fields. It is important to carefully consider the social ramifications of machine learning systems as well as fairness, transparency, privacy, and responsibility. We may work to create a future in which these technologies are helpful, responsible, and consistent with human values by proactively integrating ethical considerations throughout the development and deployment of machine learning algorithms.

1.8 Machine Learning Tools and Frameworks

Machine learning (ML) has become an innovative technology with several applications in a variety of fields. Developers and data scientists rely on reliable and effective tools and frameworks to fully utilise the power of ML. We'll look at some of the most well-liked machine learning tools and frameworks in this chapter, their capabilities, and how they make it easier to create and use ML models.

A. Tensor Flow: Google created the open-source machine learning framework Tensor Flow. It provides a complete ecosystem for the creation and deployment of ML models. Tensor Flow offers a flexible environment for numerical computing and supports both deep learning and conventional ML models. Kera's high-level API makes the process of creating neural networks easier. Tools for distributed training, visualization, and deployment across several platforms are also available with Tensor Flow.

Table 1.15: Key Features of TensorFlow

Feature	Description
Flexibility	Supports various types of ML models, including deep learning models
Scalability	Provides distributed training capabilities for large-scale models
Visualization	Offers tools for visualizing and monitoring model training
Deployment	Allows deployment on different platforms, including mobile and web

B. PyTorch:

Created by Facebook's AI Research team, PyTorch is a well-known open-source machine learning framework. It is frequently used to create deep learning models and provides dynamic computational graphs that make model training and debugging effective. The Python environment is smoothly integrated into PyTorch, which offers a Pythonic interface. Additionally, it supports distributed training and provides a wide range of tools for model deployment and visualisation.

Table 1.16: Key Features of PyTorch

Feature	Description
Dynamic graphs	Enables efficient model training and debugging with dynamic graphs
Pythonic	Offers a Pythonic interface, making it easy to integrate with Python
Distributed	Supports distributed training for large-scale models
Visualization	Provides tools for visualizing and analysing model performance

C. Scikit-learn: Scikit-learn is a well-known Python machine learning package that is open-source. It offers a comprehensive collection of tools for feature selection, model validation, and data preprocessing. Numerous machine learning (ML) algorithms, such as regression, classification, clustering, and dimensionality reduction, are supported by Scikit-learn. It provides a standard API and makes it simple to test out various models and gauge their effectiveness.

Table 1.17: Key Features of Scikit-learn

Feature	Description
Extensive API	Provides a rich set of tools for data preprocessing and model evaluation
Algorithm	Supports a wide range of ML algorithms for different tasks
Easy integration	Integrates well with other Python libraries and frameworks
Consistent API	Offers a consistent API for building and evaluating ML models

D. Microsoft Azure Machine Learning:

Microsoft offers Azure Machine Learning (AML), a cloud-based machine learning platform. For the creation and deployment of ML models, it provides a complete collection of tools and services. AML enables automated ML, which automates the selection of models and hyper parameter tuning, and offers a drag-and-drop interface for building ML pipelines. In order to facilitate development and deployment, it also interacts with well-known ML frameworks like Tensor Flow and PyTorch.

Table 1.18: Key Features of Azure Machine Learning

Feature	Description
Cloud-based	Provides a cloud-based platform for ML model development and deployment
Automated ML	Supports automated model selection and hyper parameter tuning
ML pipelines	Enables the creation of end-to-end ML pipelines with a visual interface
Framework integration	Integrates with popular ML frameworks like Tensor Flow and PyTorch

We looked at some of the well-liked machine learning tools and frameworks in this chapter. For deep learning, Tensor Flow and PyTorch are frequently used, whereas Scikit-learn offers a large range of tools for conventional ML models. Offering a cloud-based platform with integration and automation features is Microsoft Azure Machine Learning. Because each tool and framework offers a distinct set of functionality, developers and data scientists can select the one that best suits their own needs. These frameworks and tools can help ML practitioners create and deploy effective ML models faster for a variety of applications.

1.9 Machine Learning in Software Engineering

A potent tool in many fields, including software engineering, is machine learning. It has the ability to speed up development procedures, better software quality, and automate difficult operations. This chapter examines how machine learning is used in software engineering, emphasizing its advantages, difficulties, and useful applications.

Chapter 2

Machine Learning in Mathematics

Introduction

Machine learning (ML) has revolutionized various fields, including mathematics, by providing powerful tools for pattern recognition, optimization, and predictive modeling. As an intersection of computer science, statistics, and mathematical theory, machine learning utilizes algorithms that enable computers to learn from data and make decisions without explicit programming.

2.1 Role of Mathematics in ML

In mathematics, ML techniques are used for solving complex equations, proving theorems, discovering new mathematical structures, and optimizing computations. From numerical analysis to combinatory and algebraic geometry, machine learning enhances traditional mathematical approaches by automating problem-solving, finding approximations, and even generating new conjectures.

This introduction will explore key areas where machine learning contributes to mathematics, covering topics such as:

Pattern Recognition in Mathematical Data: Identifying trends and structures within large datasets.

Optimization and Numerical Methods: Using ML to improve efficiency in solving equations and optimization problems.

Automated Theorem Proving: Assisting in the discovery and verification of mathematical proofs.

Symbolic Computation and AI-assisted Problem Solving: Applying deep learning and reinforcement learning in algebra and calculus.

By integrating machine learning into mathematics, researchers and practitioners can push the boundaries of discovery, making mathematical exploration more efficient and innovative. This text aims to provide an overview of how ML is transforming the field and what the future may hold for this synergy between computation and pure mathematics.

2.2 Linear Algebra: The language of Machine Learning

Vectors and Matrices are useful in Fundamental structures for data representation. Matrix Operations such as addition, multiplication, and inversion used in transformations and optimizations. Whereas eigenvalues and eigenvectors prove to be essential for Principal Component Analysis (PCA) and dimensionality reduction. Singular Value Decomposition (SVD) Used in compression, recommendation systems, and noise reduction. Tensors and Multidimensional Data Representations: Crucial for deep learning and neural networks.

Introduction

Vectors and matrices are essential mathematical structures used in machine learning, computer science, physics, and various real-life applications. They provide a structured way to store, manipulate, and analyze data efficiently. In this discussion, we will explore their importance in data representation and real-world applications.

A. Vectors in Data Representation

Vectors are widely used to represent various types of information, such as:

In the domain of Text Data Representation like Natural Language Processing (NLP), words are represented as vectors using models like Word2Vec, GloVe, or BERT. Vector embedding are a technique in natural language processing (NLP) used to represent words or phrases as continuous numerical vectors in an abstract space.

1. Each word is mapped to a numerical vector, positioning it in a multi-dimensional space.
2. Words with similar meanings are placed closer together, allowing embeddings to encode semantic relationships.
3. These embeddings are generated using neural networks trained on large text datasets, optimizing vectors based on prediction tasks.

4. Popular embedding methods include Word2Vec and GloVe—Word2Vec utilizes a shallow neural network, while GloVe applies matrix factorization.

5. The resulting word vectors capture both syntactic and semantic properties, enabling operations like vector arithmetic. For instance, vector ("Father") - vector("Man") + vector("Woman") results in a vector close to "Mother."

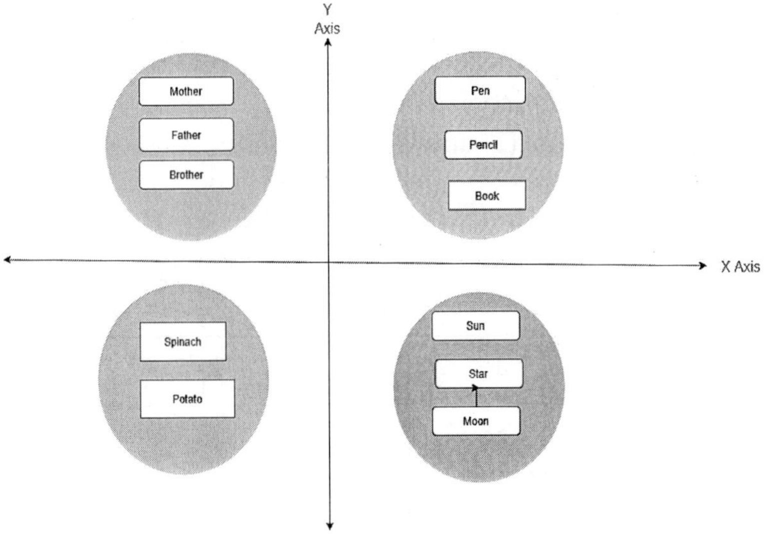

6. Embedding's help models understand analogies and generalize language patterns by associating similar words with comparable vector representations.

7. Vector dimensions typically range from 50 to 1000—higher dimensions capture more details but require greater computational resources.

B. Matrices in representation of Data

Datasets: A dataset with m samples and n features is often stored as an m×n matrix, where each row represents a data point and each column corresponds to a feature.

Images: Images are stored as matrices (or tensors) where pixel values define intensity across color channels (RGB).

Let us see examples of datasets represented as m×n matrices, where:
Where m = number of samples (data points, observations) and
n = number of features (attributes, variables)

Table 2.1: Types of representation of data

Data can be represented in two different ways	
Tabular Data	Time Series Data

1. **Tabular Data (Structured Data)**

Example: Patient Data (5 samples, 3 features)

Table 2.2: Tabular Data (Structured Data)

Patient ID	Age	Weight	Income
1	40	75	45000
2	81	59	5000
3	34	66	36000
4	57	54	78000
5	73	60	19000

Where matrix is presented as

$$C = \begin{bmatrix} 40 & 75 & 45000 \\ 81 & 59 & 50000 \\ 34 & 66 & 36000 \\ 57 & 54 & 78000 \\ 73 & 60 & 19000 \end{bmatrix}$$

2. Time Series Data (Stock Prices, IOT Sensors)

Rows represent **timestamps**, and columns represent **measured features**.

Example: Stock Price Data (5 Days, 3 Features)

Table 2.3: Time Series Data

Date	Open Price (Rs.)	Close Price (Rs.)	Volume
12-02-2024	150	155	10
13-02-2024	155	158	12
14-02-2024	158	157	8
15-02-2024	157	160	9
16-02-2024	160	162	11

$$B = \begin{bmatrix} 150 & 155 & 10 \\ 155 & 158 & 12 \\ 158 & 157 & 8 \\ 157 & 160 & 9 \\ 160 & 162 & 11 \end{bmatrix}$$

3. Tensors and Multidimensional Data Representations: Crucial for deep learning and neural networks.

Tensors are fundamental data structures in deep learning, extending the concept of matrices to higher dimensions. They allow neural networks to efficiently process large amounts of structured data, such as images, videos, and time-series signals. Tensors Crucial in Deep Learning as mentioned below:

Efficient Computation: Optimized for parallel processing on GPUs. Flexible Data Handling: Handles various data types (images, text, videos, etc.).

Supports Automatic Differentiation: Enables gradient calculations in neural networks.

Scalability: Can handle massive datasets required for AI training.

A tensor is a multidimensional array that generalizes scalars, vectors, and matrices to higher dimensions.

Table 2.4: Dimension of Tensor

Dimension of Tensor	Nomenclature	Example
zero	Scaler-A single numerical value	x=5
one	Vector-a row matrix having only one row	v=[3,7,2]
Two	Matrix	$A = \begin{bmatrix} 1 & 0 & 2 \\ 3 & 4 & 5 \\ 1 & 7 & 8 \end{bmatrix}$
Three	a stack of 2D matrices	RGB images-The height and width represent the spatial dimensions (rows and columns).The depth (3 channels) represents the Red, Green, and Blue color channels.
Four	A batch of images (batch_size × height × width × channels)	Suppose we have a batch of 2 RGB images, each with a size of 3x3 pixels: **Batch size** → 2 images **Height** → 3 pixels **Width** → 3 pixels

		Channels → 3 (Red, Green, Blue) shape of the tensor:(2,3,3,3)
Five	Dimensions: (batch size, frames, height, width, channels)	a batch of 2 videos, each containing 5 frames of 64x64 pixel RGB images shape of tensor: (2,5,64,64,3)

A. Tensors in Deep Learning

Deep learning models operate on tensors to represent and transform data. It is applicable in image processing (Computer Vision) particularly in RGB image representation. Three dimensional tensors of Shape: (Height × Width × Channels) are used here.

For example: A 256×256 color image with 3 channels (RGB) → Tensor shape: (256, 256, 3)

Also four dimensional tensors of Shape: (Batch_Size × Height × Width × Channels) are applied in a batch of images.

For example: A batch of 32 images, each 256×256 pixels with 3 channels → Tensor shape: (32, 256, 256, 3)

B. Tensors in Natural Language Processing (NLP)

Text data is represented as sequences of word embedding using 3D tensors. Three dimensional tensors of shape: (Batch Size × Sequence Length × Embedding Dimension)

For example: A batch of 64 sentences, each having 50 words, with 300-dimensional embedding → Tensor shape: (64, 50, 300)

C. Tensors in Time-Series & Sensor Data

Tensors of Shape: (Batch Size × Time Steps × Features) used in forecasting, finance, and IoT applications.

For example: A dataset with 100 time-series samples, each containing 10 time steps and 5 feature

2.3 Probability and Statistics: Managing Uncertainty in Machine Learning

Probability and statistics form the foundation of machine learning (ML) by providing methods to quantify uncertainty, make predictions, and analyze data distributions. ML models rely on these concepts to generalize from data, optimize performance, and assess reliability. Few real-world applications include Predictive Analytics: Weather forecasting, stock market prediction. Anomaly Detection: Fraud detection, network security monitoring. Natural Language Processing (NLP): Probabilistic language models (e.g., GPT, BERT).

Role of Probability in Machine Learning It consists of various applications as mentioned below.

A. Probability Distributions

Probability distributions describe how likely different outcomes are in a dataset. Common distributions used in ML include the distributions mentioned in table 2.5

Table 2.5: Types of Distribution

	Type			
Applications	Normal (Gaussian) Distribution	Bernoulli & Binomial Distributions	Poisson Distribution	Multinomial Distribution
	linear regression and neural networks.	binary classification (e.g., coin toss, spam detection)	Models rare event occurrences (e.g., number of website visits per hour).	Used in NLP for word prediction (e.g., predicting the next word in a sentence).

Example:

A model predicting the probability of rain uses a probability distribution where:

P(rain=0)= 0.7 (No rain) and P(rain=1)= 0.3 (Rain)

B.Bayes' Theorem and Conditional Probability

Bayes' Theorem is fundamental in probabilistic models like Naïve Bayes Classifier. The formula is

$$P(A/B) = P(B \mid A)P(A)/P(B)$$

It is used in spam filtering:

P (Spam|Word "free") estimates how likely an email is spam given it contains the word "free."

C.Maximum Likelihood Estimation (MLE) & Maximum A Posteriori (MAP)

MLE: Finds the parameters that maximize the probability of observed data.

MAP: Incorporates prior knowledge (Bayesian approach) for better estimation.

Role of Statistics in Machine Learning

A. Descriptive Statistics

It is applicable in summarizing and analyzing data. Mean, median, and mode are measures of central tendency that indicate the typical value in a dataset. Variance and standard deviation measure how spread out the data is around the mean. Correlation and covariance show the relationship between two variables, indicating how they change together.

Example: In feature engineering, high-variance features may indicate informative variables, while low-variance features might be redundant.

Table 2.6: Types of Measures of Central tendency

Measure	Best for	Example Use Cases
Mean	Normally distributed data	Feature scaling, linear regression, recommender systems
Median	Skewed data or presence of outliers	Salary prediction, anomaly detection
Mode	Categorical data	Finding common values, missing data imputation

B. Inferential Statistics

Help draw conclusions about populations based on sample data:

1. **Hypothesis Testing**: Used to determine statistical significance in ML models.
2. **Confidence Intervals**: Provide an estimated range for model predictions.

Example: A/B testing in online experiments uses inferential statistics to determine if a new website design improves user engagement.

C. Over fitting & Under fitting Analysis

1. **Overfitting Analysis:** Overfitting happens when a model learns **too much** from the training data, including noise and outliers. It performs very well on the training set but poorly on new, unseen data (test data).

Example: A polynomial regression model using a very high-degree polynomial for a simple linear pattern will likely over fit.

2. **Under fitting Analysis:** Under fitting happens when a model is too simple to capture the underlying patterns of the data. It performs poorly on both the training and test datasets.

Example:
Trying to fit a straight line (linear regression) to data that follows a curve will result in under fitting.

The bias-variance trade-off is a fundamental concept in machine learning that involves finding the right balance between a model's complexity and its ability to generalize to new data. High bias indicates a model that is too simple and under fits the data, while high variance suggests a model that is too complex and overfits. Achieving an optimal trade-off helps build models that perform well on both training and unseen data. Cross-validation, particularly techniques like k-fold cross-validation, is a method used to evaluate and improve model robustness. By dividing the dataset into multiple subsets and training the model on different combinations of these subsets, cross-validation provides a more reliable estimate of model performance and helps in selecting models with better generalization capabilities.

Visual Representation

When plotting training and validation loss during model training, the pattern of the curves can reveal how well the model is learning. In cases of overfitting, the training loss decreases while the validation loss starts to increase, indicating that the model is memorizing the training data but failing to generalize to new data. In underfitting, both training and validation losses remain high, showing that the model is too simple or not trained enough to capture the underlying patterns. A good fit is achieved when both training and validation losses decrease and eventually stabilize at low values, suggesting that the model is learning effectively and generalizing well to unseen data.

D. Markov Chains and Hidden Markov Models (HMMs) Used in sequential decision-making and speech recognition.

Both Markov Chains and Hidden Markov Models are powerful tools for handling sequential data and making predictions over time. They're widely used in fields like sequential decision-making, speech recognition, natural language processing (NLP), and robotics.

1. **Markov Chains:** A Markov Chain is a stochastic (random) model that represents a system transitioning from one state to another, where the probability of moving to the next state depends only on the current state (this is known as the Markov Property). It consists of

States (S): Set of possible states (e.g., Sunny, Rainy for weather)

Transition Probabilities (P): Probability of moving from one state to another P(Xt+1/Xt)

Example (Weather Prediction):

It consists of States like Sunny (S), Cloudy (C), Rainy (R) and Transition Matrix A is

$$A = \begin{bmatrix} 0.8 & 0.2 & 0.1 \\ 0.15 & 0.6 & 0.3 \\ 0.05 & 0.2 & 0.6 \end{bmatrix}$$

If it's **Sunny** today, there's an 80% chance it'll be sunny tomorrow.

Applications:

1. Stock market predictions
2. Text generation
3. Predicting next moves in board games

2. Hidden Markov Models (HMMs)

A Hidden Markov Model extends the Markov Chain by assuming the actual state is hidden (not directly observable), but we can observe outputs (emissions) that are dependent on these hidden states.

Hidden Markov Models consists of Hidden States (S)(e.g., phonemes in speech), Observations (O): (e.g., sound signals), Transition Probabilities (A): Probability of moving from one hidden state to another, Emission Probabilities (B): Probability of observing a certain output given a hidden state, and Initial State Probabilities (π): Probability of starting in a specific state

Example (Speech Recognition):

Here hidden states are words or phonemes (Hello, How, Are, You)

Also observations are in the form of acoustic signals or features extracted from speech

The model finds the most likely sequence of words given the observed sounds using algorithms like Viterbi.

3. Applications in Sequential Decision-Making:

Robot Navigation: Predicting the next move based on the current state

Speech Recognition: Decoding speech signals into text

Financial Modelling: Predicting market states

2.4 Symbolic Computation and Automated Theorem Proving

Symbolic Computation and Automated Theorem Proving are two powerful fields in computer science and mathematics that are highly intertwined with formal methods, artificial intelligence, and computational logic.

Symbolic computation: It is also known as computer algebra, is a branch of mathematics and computer science that deals with manipulating mathematical expressions symbolically rather than numerically. This involves the use of algorithms and software systems to perform algebraic manipulations, calculus, solving equations, and other symbolic operations.

Symbolic Manipulation: Instead of working with numerical values, symbolic computation works with symbols (like variables, functions, etc.), allowing the manipulation of algebraic expressions exactly.

Table 2.7: Type of Symbolic Manipulation

Type of Symbolic Manipulation	Applications
Simplification	Reducing complex expressions to simpler forms (e.g., simplifying polynomials)
Integration and Differentiation	Symbolically solving algebraic and differential equations
Linear Algebra	Manipulating matrices and vectors symbolically
Optimization	Finding optimal solutions symbolically

Tools for Symbolic Computation: There are various tools of symbolic calculation. One of the most important is Mathematica. It is a comprehensive system for symbolic mathematics. Maple is another symbolic computation system with robust mathematical capabilities. The most simplified one is SymPy, a Python library for symbolic mathematics, widely used in academic research and teaching. However Maxima is an open-source system for symbolic computation.

Automated Theorem Proving: Automated theorem proving is a subfield of logic and artificial intelligence that involves using computer programs to prove mathematical theorems. It aims to automate the reasoning process that humans traditionally use to establish the truth of mathematical statements.

It can be done by the method known as Formal Proofs. Here theorems are expressed in a formal language (like first-order logic or higher-order logic), and the proof is a sequence of logical steps that leads from axioms to the theorem.

Another method is Reasoning Algorithms. In this particular method, the software uses logical inference techniques to derive proofs automatically. These include resolution-A rule of inference used in propositional and first-order logic, Natural Deduction-A system of logical reasoning that mimics human-style proofs. Also Sequent Calculus- system to prove theorems, Model Checking- to Verify the truth of logical formulas in finite models. It has many applications in verification of correctness of software and hardware systems , mathematical Proofs in proving new theorems. As well as Artificial Intelligence in assisting in logical reasoning to demonstrate properties of algorithms or data structures and formal Methods uses logic to specify correctness of systems, often in safety-critical applications.

Various tools for Automated Theorem Proving are Coq ,Isabelle ,HOL Light and Prolog

Connection between Symbolic Computation and Automated Theorem Proving:

Table 2.8: Symbolic Computation and Automated Theorem Proving

Type	Method
Symbolic computation	Assist in automated theorem proving by providing the ability to manipulate mathematical expressions and derive intermediate results.
Automated theorem proving	Applied to symbolic expressions, often relies on efficient symbolic manipulation to simplify and explore the space of possible proofs.

Challenges to work:

The complexity of symbolic expressions and logical formulas can grow quickly, leading to scalability issues. While significant progress has been made in automated theorem proving, fully automated systems for general mathematical proof remain challenging. Many automated theorem proves require user interaction for guiding the proof search, especially for complex or novel theorems.

2.5 Machine Learning for Numerical Analysis

Machine Learning (ML) and Numerical Analysis intersect in many ways, particularly in solving complex mathematical problems, optimizing computations, and approximating solutions for differential equations, linear algebra, and other numerical methods. Let us see how ML is applied to Numerical Analysis.

A. Solving Differential Equations

Neural Networks for PDEs and ODEs: Deep learning models, such as physics-informed neural networks (PINNs), are used to approximate solutions to partial differential equations (PDEs) and ordinary differential equations (ODEs). Instead of explicitly solving the equations with numerical methods, a neural network is trained to approximate the solution. The network takes input variables (e.g., time, spatial coordinates) and outputs an estimated solution. Unlike traditional methods, PINNs do not require spatial discretization. PINNs work well for high-dimensional PDEs, unlike grid-based methods. It can incorporate real-world data to improve solution accuracy. Once trained, the neural network can quickly provide solutions for new scenarios.

Example: Understanding blood flow in arteries is crucial for diagnosing cardiovascular diseases. The Navier-Stokes equations, which describe fluid motion, govern blood flow but are complex to solve for real patient-specific arteries. In such cases PINNs approximate the blood flow solution without needing a fine mesh, reducing computation time. Doctors can use PINN-based models to assess blood flow in real-time during surgery or diagnosis.

B.Optimization Techniques

Gradient Descent Variants: Many numerical methods, such as Newton's method, are optimized using machine learning approaches like Adam, RMS prop, and momentum-based optimizers. Gradient descent is a fundamental optimization technique used in machine learning and numerical methods to minimize functions, such as those arising in Newton's method for solving nonlinear equations. Various advanced optimizers improve upon basic gradient descent by accelerating convergence and handling complex landscapes.

Example: RMS prop normalizes gradients to prevent large updates and improve stability.

Adam (Adaptive Moment Estimation) Combines momentum and adaptive learning, making it widely used in deep learning.

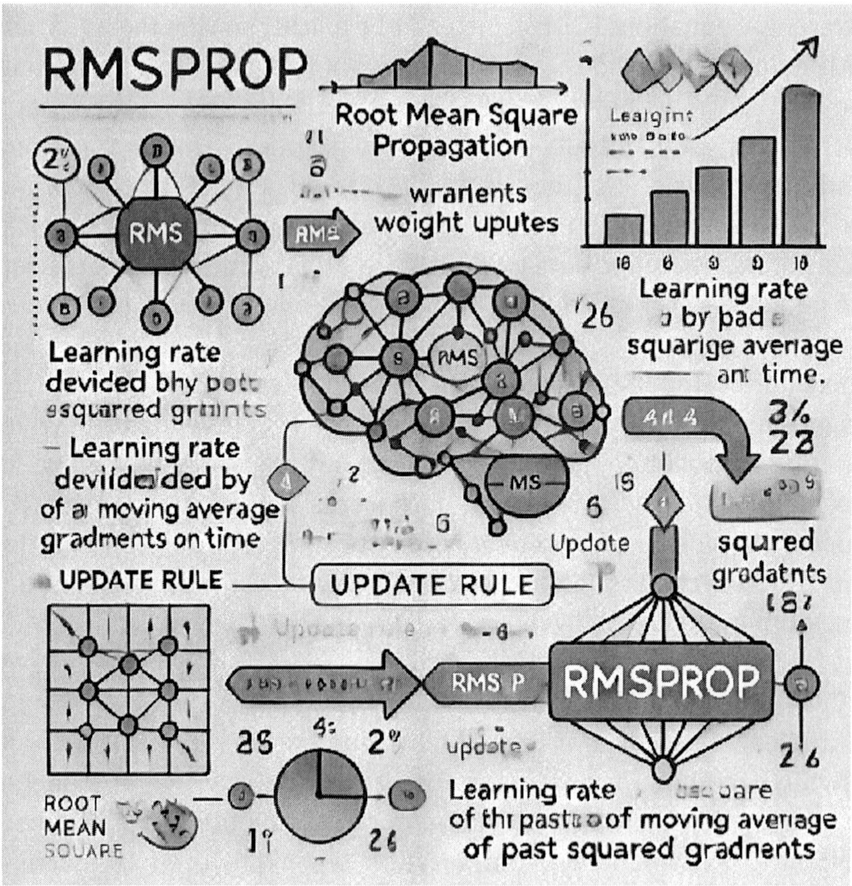

Fig: RMS Prop

3. Solving Linear Systems and Eigenvalue Problems

Graph Neural Networks (GNNs) applied in numerical linear algebra, such as predicting eigenvalues and solving graph-based numerical problems. GNNs are deep learning models designed to process graph-structured data, making them highly effective for numerical linear algebra tasks, such as predicting eigenvalues and solving graph-based numerical problems. GNNs model numerical problems as graphs, where sparse matrices are represented by nodes and edges. Through message passing, nodes exchange information to capture relationships. After several steps, a neural network predicts key numerical properties like eigenvalues and matrix factorizations. Once trained, GNNs solve problems faster than traditional methods.

Example: In financial fraud detection, identifying fraudulent transactions in banking systems is crucial.

GNNs help model financial transactions as graphs to detect unusual patterns, flagging fraudulent activities. In GNN, nodes represent bank accounts or users, edges represent transactions between them, fraudulent activity is highlighted.

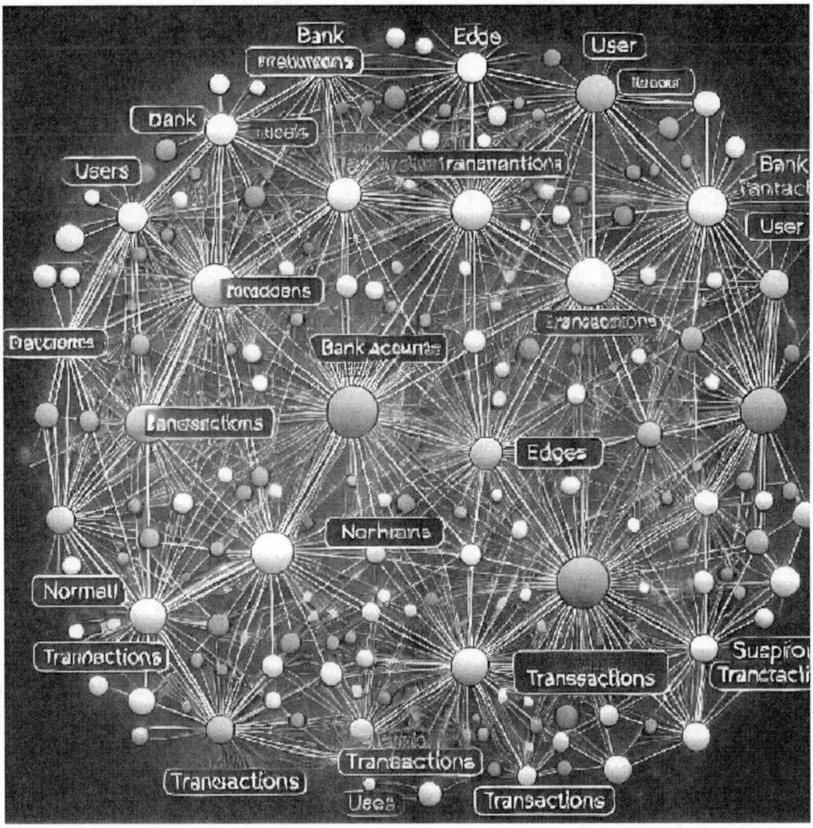

Fig: GNN assistance in financial fraud detection

4. Numerical Integration and Differentiation

Data-Driven Quadrature Rules: Machine learning techniques can optimize numerical integration methods. Traditional quadrature rules (like Simpson's rule or Gaussian quadrature) approximate integrals using predefined weight and node selections. Machine learning enhances this process by optimizing quadrature rules based on data, improving accuracy and efficiency. ML models analyze function behaviors to adaptively choose quadrature points and weights. It reduces errors in high-dimensional or irregular functions compared to fixed rules. It requires fewer function evaluations, making it ideal for complex scientific simulations.

Machine learning can optimize numerical integration methods by learning from data to create better approximations of integrals than traditional rules like Simpson's or Gaussian quadrature, based on fixed nodes and weights.

Example: Suppose you want to numerically integrate a function f(x) over the interval [a,b]. Traditional Gaussian quadrature selects specific points xi and weights wi to approximate:

$$\int_a^b f(x)dx \approx \sum_{i=1}^{n} w_i f(x_i)$$

IN the case where function f(x) has complex or irregular behavior (like sharp peaks or discontinuities), fixed rules fail to result properly.

ML-Based Approach:

Instead of using fixed xi and wi, you can **train a machine learning model** (e.g., a neural network) to learn optimal nodes and weights from a dataset of similar functions. The model adjusts the integration scheme to better fit the function class it was trained on.

Example: Approximate the integral of a function f(x) over the interval [0,1], where f(x) comes from a specific family of functions eg. $f(x) = e^{-x^2}\sin(kx)$ for different values of k.

Step1: Create a dataset of functions with known integrals computed using a very accurate method or symbolic integration.

Use a high-accuracy method (like 100-point Gauss quadrature) to compute their "true" integrals.

Step 2: To define the Model

Train a small neural network that takes in function values at a few fixed sample points (say, 5 points in [0, 1]) and predicts a weight vector

$w = [w_1, w_2, w_3, w_4, w_5]$

$$\int_0^1 f(x)dx \approx \sum_{i=1}^{5} w_i f(x_i)$$

The approximate integral becomes

Here:

Here xi are fixed sample points (e.g., equally spaced or Chebyshev nodes).

Also wi are **learned weights are predicted** by the model based on the shape of f(x).

Step 3: Train the Model

Minimize the error between the model-predicted integral and the true integral over the training set:

$$Loss = \frac{1}{N}\sum_{k=1}^{N}\left(\sum_{i=1}^{5} w_i^k f_k(x_i) - I_k^{true}\right)$$

After training, this model learns how to adapt the weights based on function features. For any new functions it can produce better approximations than fixed-rule quadrature methods like Simpson's or Gaussian quadrature.

$$f_1(x) = e^{-x}\cos(x)$$

$$f_2(x) = e^{-x}\cos(2x)$$

$$f_3(x) = e^{-x}\cos(3x)$$

....

$$f_n(x) = e^{-x}\cos(nx)$$

2.6 Pattern Recognition in Mathematical Data

Pattern recognition in mathematical data uses machine learning and statistical techniques to identify structures, trends, and relationships in numerical datasets. Feature extraction involves identifying fundamental mathematical properties such as symmetry and periodicity, which help in understanding underlying structures in data. Machine learning models, including neural networks, clustering algorithms, and symbolic regression, are then applied to detect hidden patterns within the data. These insights enable accurate prediction and classification, allowing for the analysis of mathematical sequences, equation solving, and trend forecasting.

Example:

1.In prime number distribution, Identifying patterns in prime numbers (e.g., the Riemann Hypothesis).Machine learning models analyze large datasets of prime numbers to find hidden structures in their distribution.

Q. "Can we predict the gap between consecutive prime numbers using machine learning?"

Ans: Yes we can. Generate a dataset of prime numbers (e.g., using the Sieve of Eratosthenes).

Compute the gap between consecutive primes (for prime pn the gap is pn+1−pn)

2.In fractal and Chaos theory, recognizing self-similar patterns in fractals (e.g., Mandelbrot Set). AI detects underlying mathematical rules in chaotic systems, improving weather forecasting and financial modeling.

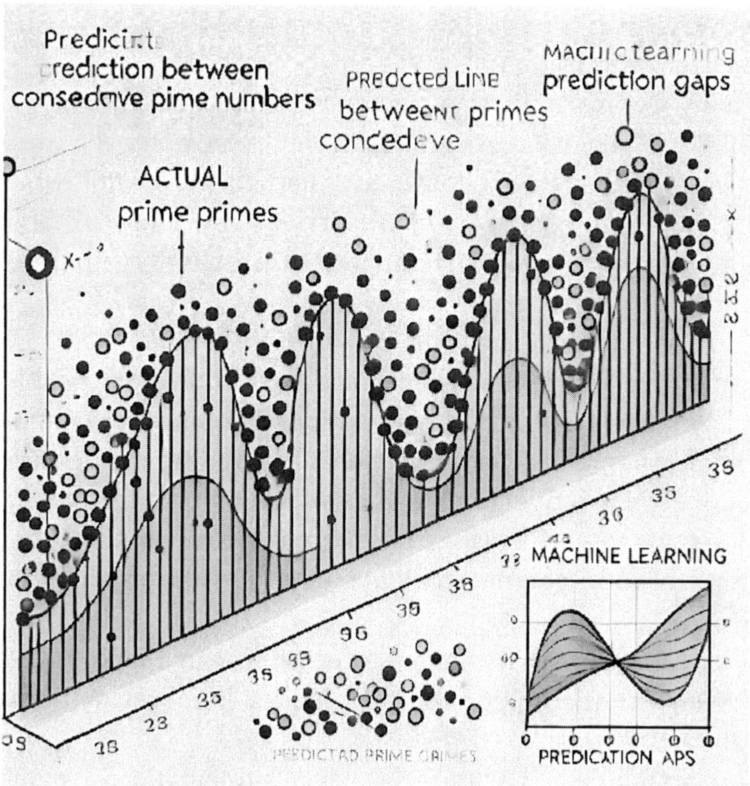

Fig: ML Prediction APS

2.7 Optimization Techniques Using ML

Machine learning enhances optimization by improving efficiency, accuracy, and adaptability in solving complex problems.

1. Gradient-Based Optimization:

Gradient Descent is an optimization algorithm used to minimize a function by iteratively updating its parameters in the direction of the negative gradient. It is widely used in training machine learning models, especially neural networks. Example: Gradient Descent and its variants (Adam, RMSprop, Momentum, AdaGrad). It Computes the gradient of the loss function and updates parameters using the particular formulae. $\omega = \omega - \eta . \nabla L(\omega)$ where ω are model parameters, η is the learning rate, and $\nabla L(\omega)$ is the gradient. ML adjusts learning rates adaptively, preventing overshooting or slow convergence in neural networks.

2. Bayesian Optimization:

Bayesian Optimization is a sequential model-based optimization (SMBO) technique used to efficiently find the optimal solution of expensive-to-evaluate functions. It is particularly useful when the function is complex, expensive, or black-box (e.g., hyperparameter tuning in deep learning).It is helpful if Gradient-based methods cannot be applied.

The number of function evaluations must be minimized.

Example: Hyper parameter tuning in deep learning models. ML Uses probability models to find optimal solutions with fewer evaluations.

Problem Statement: A company wants to optimize the performance of a deep learning model for image classification on a medical dataset (e.g., detecting tumors in MRI scans). Training the model is expensive, taking several hours per run, so traditional grid search or random search is inefficient.

Solution: Instead of trying every possible combination of hyper parameters, Bayesian Optimization predicts the best settings with minimal training runs.

Hyper parameters to Optimize considers learning rate (η), number of layers in the neural network, batch size, dropout rate

Step 1: Use a Surrogate Model

A Gaussian Process (GP) models the relationship between hyper parameters and validation accuracy.

Step 2: Select the Next Set of Hyper parameters with an Acquisition Function

The acquisition function suggests the most promising hyper parameters based on prior evaluations.

Step 3: Train the Model and Update the Surrogate Model Train the deep learning model using the suggested hyper parameters & update the GP with the new performance results. Repeat the process until the convergence. The final model achieves higher accuracy while reducing computation time by over 50%.

3. Evolutionary Algorithms:

Evolutionary Algorithms are optimization techniques inspired by natural selection. They are used to solve complex problems where traditional gradient-based methods may not work well. It initializes the population by creating a random set of possible solutions (individuals). It measures how good each solution is using a fitness function. Pick the best solutions (individuals with higher fitness) for reproduction. In recombination, it combines two parent solutions to create offspring (new solutions).

Small random changes made by introducing small variations in offspring to maintain diversity.

Repeat procedure for multiple generations resulting over time, the Population evolves, and better solutions emerge.

Table 2.9: Types of Evolutionary Algorithms

Types of Evolutionary Algorithms			
Genetic Algorithms	**Genetic Programming**	**Evolution Strategies**	**Particle Swarm Optimization**

Used for optimization problems	Evolves programs instead of numeric solutions.	Focuses on self-adaptation and mutation-based optimization.	Models optimization as a swarm of moving particles.

2.8 Applications of fixed point theory in machine learning

Fixed point theory has several interesting and valuable applications in machine learning, particularly in understanding the behavior of iterative algorithms, stability, and convergence. Let us see a few key applications.

A. Training Algorithms (Convergence Analysis)

Many machine learning algorithms are iterative by nature. Fixed point theory provides a framework for analyzing whether these algorithms will converge to a stable point.

Example: Gradient descent and its variants can often be viewed as fixed point iterations of a mapping T(x), where the update rule is xk+1=T(xk)

Role of Fixed Points: A point x such that T(x)=x corresponds to a stable solution of the learning problem (e.g., a local minimum of the loss function).

B. Neural Networks

Activation Functions: Some training methods for deep neural networks can be analyzed as finding a fixed point in a function space, particularly in recursive architectures. Equilibrium Models: Deep Equilibrium Models (DEQs) directly define the output of a neural network as the fixed point of a system of equations, removing the need for traditional layered architectures.

C. Game Theory and Multi-Agent Systems

Fixed point theorems such as Brouwer's and Kakutani's are used to prove the existence of Nash equilibria in multi-agent environments, which is critical in machine learning settings involving competition or collaboration between agents (e.g., generative adversarial networks)

D. Generative Models

GANs: Training a GAN can be seen as a dynamical system where both generator and discriminator are trying to reach an equilibrium—effectively a fixed point in the parameter space.

E. Meta-learning and Hyper parameter Optimization

Some meta-learning procedures involve solving bi-level optimization problems. The inner optimization can often be seen as reaching a fixed point of a certain operator, which influences the outer optimization loop.

Chapter 3

Machine Learning in Physics

3.1 Introduction

Machine learning (ML) is transforming many fields of science, including physics. It provides new tools for solving complex problems, analyzing large datasets, and making predictions in areas that are difficult or impossible to model using traditional methods. By leveraging algorithms and statistical models, machine learning enables researchers to extract valuable insights from data, discover patterns, and even formulate physical laws that govern natural phenomena. In physics, this intersection of data science and theoretical understanding has led to a variety of breakthroughs across various domains, such as quantum mechanics, fluid dynamics, astrophysics, and more.

Physics involves highly complex systems, often characterized by a large number of variables, intricate interactions, and nonlinear behavior. Traditional techniques, such as analytical models or numerical simulations, can struggle with systems that have a high degree of complexity or when large-scale data is involved. ML methods excel in these scenarios, offering the following advantages:

- **Handling Large Data Sets:** ML can process and analyze vast amounts of data that would be impractical for traditional methods.
- **Pattern Recognition:** ML algorithms are designed to identify hidden patterns and correlations in datasets, even when the underlying structure is not explicitly known.
- **Prediction and Forecasting:** Once trained on sufficient data, ML models can make accurate predictions about the future behavior of a system.
- **Uncertainty Quantification:** ML methods can help quantify uncertainty and make decisions based on incomplete or noisy data.

A. Impact of machine learning in physics:

- **Quantum Mechanics and Quantum Computing:** Machine learning has been applied to problems in quantum mechanics, such as quantum state tomography, predicting quantum system behavior, and optimizing quantum algorithms. It is also crucial in the development of quantum computing, where ML models assist in error correction and the optimization of quantum gates.

- **Astrophysics and Cosmology:** ML is widely used to analyze large datasets from telescopes, such as those capturing information about galaxies, stars, and black holes. Machine learning techniques like supervised learning and deep learning are used to classify astronomical objects, detect gravitational waves, and understand the cosmic microwave background radiation.
- **Condensed Matter Physics:** ML is useful in condensed matter physics to analyze complex materials, phase transitions, and quantum states. For instance, neural networks are employed to identify patterns in data and simulate the behavior of particles in a system. ML has also been instrumental in identifying new materials with specific properties for use in technology.
- **Particle Physics:** In high-energy physics, such as research done at particle accelerators (e.g., CERN), ML is employed for tasks like event classification, anomaly detection, and particle track reconstruction. It can help analyze the massive amounts of data generated during particle collisions, enabling researchers to detect rare phenomena and new particles.
- **Fluid Dynamics and Plasma Physics:** Fluid dynamics is often governed by the Navier-Stokes equations, which are difficult to solve analytically for most practical situations. ML techniques, such as neural networks, can approximate solutions to these equations or learn from data to make accurate predictions about fluid flow, turbulence, and plasma behavior.
- **Climate Modeling and Weather Forecasting:** Machine learning is increasingly used to improve climate models and weather predictions by identifying patterns in climate data that traditional models might miss. It can also be applied to study phenomena like ocean currents, atmospheric conditions, and the impact of human activity on the environment.

B. Challenges and Future Directions

While the integration of machine learning with physics offers significant potential, there are several challenges that need to be addressed:

- **Interpretability:** ML models, especially deep learning models, can be seen as "black boxes" that do not offer easy interpretation of the physical phenomena they model. Physicists are working on methods to make these models more transparent and interpretable.
- **Data Quality and Availability:** In many cases, high-quality labeled data can be scarce, especially in experimental physics. Ensuring accurate and reliable data is crucial for the success of ML models.
- **Scalability:** Some machine learning methods may struggle with the sheer scale of data encountered in large-scale experiments or simulations. Developing methods that can efficiently handle vast datasets is a key area of research.
- **Integration with Physical Theories:** While machine learning can uncover patterns and make predictions, integrating these insights with well-established physical theories remains a complex task. There is a growing interest in developing hybrid models that combine ML with traditional physics principles.

Development of Quantum Computing with Machine Learning Assistance

Development Area	Role of ML Models	Impact on Quantum Computing
Quantum Error Correction	Detects and corrects quantum errors using reinforcement learning and pattern recognition.	Enhances fault tolerance, reducing decoherence and noise in quantum circuits.

Optimization of Quantum Gates	Optimizes gate sequences using ML-driven algorithms to minimize gate errors and execution time.	Improves computational efficiency and fidelity of quantum operations.
Quantum State Prediction	Uses deep learning to predict quantum states and their evolution over time.	Enables better qubit control, improving quantum simulations and algorithms.
Noise Reduction & Mitigation	Identifies and filters out noise in quantum computations through supervised and unsupervised learning.	Enhances stability and accuracy of quantum computations.
Quantum Hardware Calibration	Automates hardware tuning using ML techniques like Bayesian optimization.	Reduces manual intervention, improving performance and scalability of quantum processors.
Hybrid Quantum-Classical Algorithms	ML models assist in bridging classical and quantum computations for complex problem-solving.	Accelerates applications in cryptography, materials science, and AI.
Quantum Circuit Compilation	Optimizes circuit layouts and gate placements using ML heuristics.	Lowers resource requirements, making quantum programs more efficient.

C. Data-Driven Discoveries in Physics

Machine learning (ML) has become a transformative tool in physics, enabling data-driven discoveries that were previously not possible or were computationally prohibitive. By leveraging ML algorithms, physicists can analyze large datasets, uncover hidden patterns, and make predictions that guide experimental and theoretical advances. The rise of ML has accelerated discoveries across various fields of physics, from particle physics to quantum mechanics and astrophysics. Below, we explore how data-driven discoveries are being made in physics through the application of machine learning.

1. Automated Discovery of Physical Laws and Models

Traditionally, physics discoveries often involved deriving theoretical models to explain observed phenomena. ML has introduced the possibility of "data-driven" model discovery, where algorithms attempt to identify underlying physical laws directly from data without explicit human intervention.

- **Symbolic Regression:** One of the key techniques in data-driven physics is symbolic regression, which uses genetic algorithms to search for mathematical expressions that fit a given dataset. This method has been used to discover new mathematical relations in physical systems. For example, symbolic regression has been used to uncover simplified forms of complex physical equations or discover new relationships in fluid dynamics or celestial mechanics.
- **Example:** In fluid dynamics, researchers have used symbolic regression to rediscover well-known equations such as the Navier-Stokes equations for turbulence, as well as to find new formulas for flow dynamics.
- **Discovering New Laws of Motion or Conservation:** ML models can also be trained to discover new laws or conservation principles by analyzing experimental data. A notable example is the use of machine learning to propose alternative theories of motion or new symmetries in classical and quantum systems.

2. Discovery of New Materials in Condensed Matter Physics

Condensed matter physics has been deeply impacted by machine learning, particularly in the discovery of new materials with unique properties.

- **Materials Discovery:** Traditional methods for discovering new materials involve trial and error or physics-based simulations, which are often computationally expensive. Machine learning accelerates this process by predicting the properties of materials based on their structure, composition, and other features.
- **Example:** ML models have been used to predict superconducting materials, which are materials that conduct electricity with zero resistance at high temperatures. By analyzing thousands of candidate materials, ML models can predict which materials have the potential to exhibit superconductivity, reducing the time and resources required for experimental verification.
- **Quantum Materials:** ML has also been applied to the discovery of quantum materials with exotic properties, such as topological insulators, which are materials that have special conductive properties on their surface but are insulative inside. By training models on experimental data, ML can help identify promising materials with these unique quantum behaviors.

3. Astrophysics: Exoplanet Detection and Cosmological Insights

Astrophysics is another area where machine learning has significantly enhanced our ability to analyze data and make new discoveries, particularly with the advent of large-scale surveys and telescopes.

- **Exoplanet Detection:** One of the most exciting applications of ML in astrophysics is the detection of exoplanets—planets outside our solar system. Traditional methods for detecting exoplanets, such as the transit method (detecting the dimming of stars as planets pass in front of them) or radial velocity (measuring the star's wobble due to gravitational pull from orbiting planets), generate large amounts of noisy data. ML algorithms, especially deep learning models, have been applied to these datasets to automate the identification of exoplanet signals.

• **Example:** NASA's Kepler space telescope, which has discovered thousands of exoplanets, uses machine learning algorithms to identify patterns in the light curves of stars. The algorithms have drastically improved the efficiency of exoplanet detection.

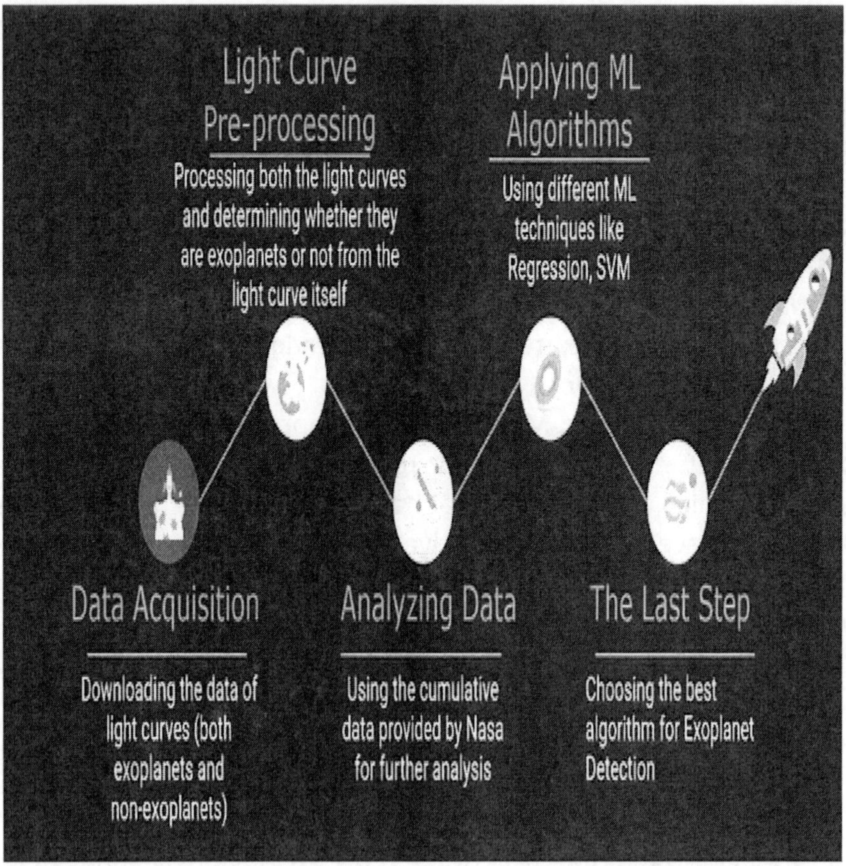

• **Gravitational Wave Detection:** The detection of gravitational waves, ripples in spacetime caused by catastrophic astrophysical events like black hole mergers, has also benefited from machine learning. ML models are used to filter and analyze data from observatories like LIGO (Laser Interferometer Gravitational-Wave Observatory) to detect subtle gravitational wave signals in noisy data.

• **Cosmological Parameter Estimation:** ML is used to estimate cosmological parameters, such as the Hubble constant (which describes the rate of expansion of the universe), from galaxy surveys and large-scale structure data. Machine learning techniques like Gaussian Processes or neural networks help in constraining these parameters more accurately and efficiently than traditional methods.

4. Particle Physics: Anomaly Detection and New Particles

In high-energy physics, particularly at large particle accelerators like CERN's Large Hadron Collider (LHC), ML is playing a pivotal role in detecting new particles, classifying events, and finding rare phenomena that would otherwise be lost in vast datasets.

- **Anomaly Detection in Particle Collisions:** High-energy particle collisions generate enormous amounts of data. Machine learning algorithms, such as deep neural networks, are used to analyze these collisions to detect anomalies or unexpected events that may signal new particles or unknown physics. This approach is particularly useful for identifying rare events like the Higgs boson or new forms of matter.

  Example: The discovery of the Higgs boson in 2012 was facilitated by ML techniques to sift through data and identify possible signals from particle collisions.

- **Jets and Particle Classification:** ML algorithms are used to classify different types of particles produced in particle collisions. Jets—streams of particles that result from quarks and gluons produced in high-energy collisions—are classified using ML models to understand the fundamental interactions between particles and to search for new particles.

5. Quantum Mechanics: Improving Simulations and Quantum State Reconstruction

Machine learning has found applications in quantum mechanics, particularly in areas where traditional computational techniques face significant challenges, such as simulating quantum systems or reconstructing quantum states from experimental data.

- **Quantum State Tomography:** Quantum state tomography is the process of reconstructing the quantum state of a system from measurements. ML models, particularly deep learning networks, have been applied to quantum state tomography to improve the accuracy and efficiency of state reconstruction from limited and noisy data.

- **Quantum Circuit Design:** In quantum computing, designing efficient quantum circuits is a complex task. Machine learning can optimize the design of quantum circuits by training models on existing quantum circuits to suggest more efficient configurations. Reinforcement learning has been used to optimize quantum gate operations, improving quantum computation processes.

6. Fluid Dynamics and Turbulence

Fluid dynamics, especially the study of turbulence, has long been a challenging field for both theoretical and computational physics due to the complexity of the governing equations. Machine learning is now providing new insights into turbulence, enabling faster simulations and better predictions.

- **Turbulence Prediction:** ML models, particularly neural networks, have been trained on experimental data of turbulent flows to predict the behavior of fluids in complex situations. This can potentially lead to more accurate and less computationally expensive simulations of turbulence in engineering and natural systems.
- **Learning from Simulations:** ML is also used to analyze the results of high-resolution fluid simulations to identify dominant features or patterns that could lead to simplified models or better understand the underlying dynamics of fluid motion.

3.2 Quantum Computing and Machine Learning: A Synergistic Relationship

Quantum computing and machine learning (ML) are two fields that are rapidly evolving and have the potential to revolutionize many areas of science, technology, and industry. When combined, quantum computing could drastically enhance machine learning algorithms, particularly in dealing with large datasets, complex computations, and optimization problems. Conversely, machine learning can aid in the development of quantum algorithms, help overcome challenges in quantum computing, and optimize quantum systems. This symbiotic relationship could unlock new possibilities for solving previously intractable problems.

A. Quantum Computing: An Overview

Quantum computing is a revolutionary approach to computation that leverages the principles of quantum mechanics, such as superposition, entanglement, and quantum interference, to perform computations. Unlike classical computers that process information in binary form (0s and 1s), quantum computers use quantum bits or qubits, which can exist in multiple states simultaneously (superposition). This allows quantum computers to process vast amounts of data in parallel, offering the potential for exponential speedups in specific problem areas.

B. Key features of quantum computing include:

- **Superposition:** A qubit can represent multiple states at once (e.g., both 0 and 1).
- **Entanglement:** Qubits can become correlated in ways that allow for the transfer of information between qubits, even over large distances.
- **Quantum Interference:** Quantum algorithms can manipulate the interference of qubit states to amplify correct answers and diminish incorrect ones.
- These features make quantum computers particularly powerful for tasks involving complex calculations, such as optimization, cryptography, and simulation of quantum systems.

C. Machine Learning: An Overview

Machine learning is a subset of artificial intelligence (AI) where algorithms learn from data to recognize patterns, make decisions, and improve over time. There are various types of ML, including supervised learning, unsupervised learning, and reinforcement learning. ML models are used in diverse fields, from image recognition and natural language processing (NLP) to medical diagnostics and autonomous vehicles.

Machine learning algorithms can process and analyze large datasets to:

- Identify patterns in data.
- Make predictions based on past data.
- Optimize solutions for complex problems.

The effectiveness of machine learning algorithms often depends on the availability of sufficient data, computational power, and optimization techniques.

D. The Synergy Between Quantum Computing and Machine Learning

The intersection of quantum computing and machine learning opens up new avenues for solving complex problems more efficiently. Quantum computing has the potential to address key challenges in machine learning, such as computational bottlenecks, optimization problems, and the scalability of algorithms.

Here are some ways quantum computing can enhance machine learning:

1. Quantum Speedup for Machine Learning Algorithms

Classical machine learning algorithms can be slow and computationally expensive, particularly when processing large datasets. Quantum algorithms have the potential to speed up key processes in ML, such as matrix inversion, linear algebra, and data classification.

- **Quantum Linear Algebra:** Quantum computers can accelerate matrix multiplication and other linear algebra operations, which are central to many machine learning algorithms, such as principal component analysis (PCA), support vector machines (SVMs), and neural networks. Quantum algorithms like Harrow-Hassidim-Lloyd (HHL) offer exponential speedup for solving systems of linear equations.
- **Quantum Fourier Transform (QFT):** QFT, a quantum analog of the classical Fourier transform, is a core component of algorithms like Shor's algorithm. It has applications in quantum machine learning, especially in spectral analysis and signal processing.

2. Quantum Machine Learning (QML) Algorithms

Quantum machine learning (QML) is an emerging field that combines quantum computing with traditional machine learning techniques. Several quantum algorithms have been proposed that can potentially outperform classical ML algorithms, especially for high-dimensional data and large-scale problems.

- **Quantum Support Vector Machines (QSVMs):** Support vector machines are a powerful tool for classification problems. A quantum version of SVM can speed up training and improve performance by leveraging quantum data encoding and quantum-enhanced kernel functions.
- **Quantum Neural Networks (QNNs):** Neural networks are the backbone of deep learning. QNNs use quantum circuits to represent and process data in quantum states. Quantum neural networks could provide exponential speedups for training deep neural networks by exploiting quantum parallelism.
- **Quantum k-Means Clustering:** Quantum algorithms for clustering problems, such as k-means clustering, can be more efficient by using quantum superposition to evaluate all cluster centroids simultaneously, reducing the time complexity compared to classical approaches.
- **Quantum Boltzmann Machines (QBM):** Boltzmann machines are generative stochastic neural networks. Quantum Boltzmann machines leverage quantum systems to efficiently simulate the states of these networks, potentially accelerating learning and enhancing the ability to model complex data distributions.

E. Quantum Advantage in Data Optimization

Many machine learning problems involve optimization tasks, such as minimizing the error in a predictive model or finding the best parameters for a machine learning algorithm. Quantum computing can offer advantages in optimization problems that are common in machine learning.

- **Quantum Approximate Optimization Algorithm (QAOA):** The QAOA is designed to solve combinatorial optimization problems. It can be applied to ML tasks like hyperparameter optimization, feature selection, and training deep learning models.

- **Quantum Annealing:** Quantum annealing, implemented on quantum processors like D-Wave, is used for optimization tasks by exploiting quantum tunneling to find the global minimum of a function. This has applications in various ML tasks, including optimizing training parameters and solving complex combinatorial problems.

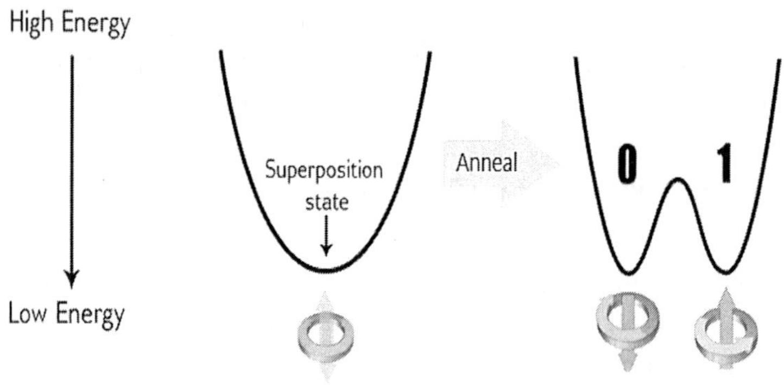

3.3.5. Data Encoding and Quantum State Representation

One of the challenges in applying quantum computing to machine learning is the efficient encoding of classical data into quantum states. Quantum machine learning algorithms rely on quantum states to represent data, and efficient encoding methods are crucial for achieving quantum speedup.

- **Quantum Feature Mapping:** Quantum algorithms can encode classical data into quantum states in such a way that quantum computers can process high-dimensional feature spaces more efficiently. This can enhance the performance of quantum classifiers and regression models.
- **Quantum State Representation for High-Dimensional Data:** Quantum computers are particularly effective at handling high-dimensional data. For example, quantum computers can process high-dimensional vectors and matrices with fewer resources, which could lead to faster training times for machine learning models.

F. Enhancing Deep Learning and Reinforcement Learning

Quantum computing holds great promise for enhancing reinforcement learning (RL) and deep learning, two of the most computationally intensive areas of machine learning.

- **Quantum-enhanced Reinforcement Learning (QRL):** In reinforcement learning, an agent learns to make decisions by interacting with its environment. Quantum reinforcement learning could improve the exploration-exploitation tradeoff and speed up the learning process by providing quantum advantages in decision-making and policy optimization.
- **Quantum Gradient Descent:** Gradient descent is a fundamental optimization algorithm used to train neural networks. Quantum versions of gradient descent algorithms could significantly speed up convergence and allow for more efficient training of deep networks.

While the potential for quantum-enhanced machine learning is enormous, there are several challenges to overcome:

- **Quantum Hardware Limitations:** Current quantum computers are still in their early stages (NISQ – Noisy Intermediate-Scale Quantum) and are limited by noise, decoherence, and error rates. This means that practical and reliable quantum machine learning algorithms are still far from being widely available.
- **Data Encoding and Quantum-Classical Interface:** Efficiently encoding classical data into quantum states and interfacing quantum algorithms with classical data systems remains a major hurdle. New techniques need to be developed to seamlessly integrate quantum computing into real-world machine learning workflows.
- **Scalability:** Many quantum algorithms, particularly those for machine learning, still face scalability issues due to the limited number of qubits available on current quantum computers.

3.3 Neural Networks in Chaptericle Physics

In **particle physics**, neural networks (NNs) and **machine learning (ML)** techniques have revolutionized how scientists process and analyze the vast amounts of data generated by high-energy physics experiments.

These neural networks enable faster, more accurate predictions, classification, and pattern recognition, helping physicists uncover new phenomena and gain insights into fundamental questions about the universe.

The use of machine learning and neural networks in particle physics typically revolves around the classification of particle collisions, the identification of rare events, optimizing the performance of detectors, and simulating complex physical systems. Neural networks, as part of the broader machine learning paradigm, are especially suited to handling the large-scale, high-dimensional data that modern particle detectors produce.

Here's an overview of the role neural networks and machine learning play in particle physics.

A. Key Applications of Neural Networks in Particle Physics

1. Event Classification and Signal Detection

Particle physics experiments, especially at large particle accelerators like the **Large Hadron Collider (LHC)**, generate a huge volume of data. In these experiments, particles collide at high energies, producing a variety of detectable events. The challenge lies in identifying interesting and potentially new physics (e.g., rare particle decays or new particles) from the overwhelming background noise of common events.

Neural networks, especially **deep neural networks (DNNs)**, are widely used to classify collision events and detect meaningful signals that deviate from normal background events.

- **Example:** The discovery of the **Higgs boson** in 2012 at the LHC relied heavily on machine learning, including neural networks, to classify particle collision events and identify those that matched the theoretical predictions of the Higgs decay signature.
- **Neural Networks in Classification:** Using labeled datasets, DNNs can be trained to recognize different types of particle interactions. By learning the characteristics of known interactions, neural networks are capable of distinguishing signal events (indicating new particles or interactions) from background noise.

2. Anomaly Detection

In particle physics, scientists are often searching for unknown or unexpected events. These rare events could point to new physics, such as the existence of **dark matter** or **supersymmetric particles**.

- **Anomaly detection with autoencoders: Autoencoders**, a type of neural network used for unsupervised learning, have been used for anomaly detection. By learning the normal patterns of data (such as typical particle collisions), an autoencoder can identify deviations from these patterns that may indicate an unknown phenomenon.
- **Example:** In high-energy physics, anomaly detection techniques can be used to uncover new particles, such as potential signals of **dark matter** candidates, by detecting deviations from the expected background events in particle collision data.

3. Particle Tracking and Reconstruction

In particle detectors, tracking is the process of identifying the trajectories of charged particles as they travel through various detector components (such as silicon trackers, calorimeters, and muon detectors). Neural networks have been used to improve particle tracking and the reconstruction of particle paths in high-energy environments.

- **Reconstructing Particle Tracks:** Neural networks, particularly **convolutional neural networks (CNNs)**, can be trained to recognize patterns in the detector's data, such as the hits in a calorimeter or tracking detector, and identify the most probable particle trajectories.
- **Example:** At the **ATLAS** and **CMS** experiments at the LHC, CNNs have been used to improve the accuracy of tracking particle trajectories, especially in complex environments where traditional algorithms might struggle due to noise or interference.

4. Simulation of Particle Collisions

Simulating particle collisions is essential for understanding how particles interact in different experimental settings. Traditional methods for simulation involve Monte Carlo (MC) techniques, but they can be computationally expensive and slow, particularly for complex models. Neural networks, especially **generative models** like **Generative Adversarial Networks (GANs)**, can be used to simulate particle collisions more efficiently.

- **Faster Simulations:** Neural networks can be trained to generate synthetic data that mimics real experimental data. This can help accelerate the simulation process and reduce the need for time-consuming Monte Carlo simulations.
- **Example:** GANs have been used to simulate realistic particle collision events, providing a faster alternative to traditional simulation methods while maintaining accuracy. These simulations can be used to validate detector performance or optimize experimental design.

5. Optimizing Detector Performance

The configuration of particle detectors is critical to collecting high-quality data. Neural networks can be applied to optimize the design and operation of detectors, improving their sensitivity and efficiency.

- **Example:** Neural networks can be used to automatically tune the parameters of particle detectors, improving the accuracy of particle tracking, energy measurements, and event reconstruction. They can also assist in the calibration of detectors, adjusting for factors like noise, drift, or other systematic errors.

6. Data Compression and Feature Selection

Particle physics experiments generate enormous amounts of data, and handling this data efficiently is crucial. Neural networks can assist in **data compression** and **feature selection**, reducing the dimensionality of the data while preserving the most relevant information.

- **Autoencoders for Compression:** Neural networks, particularly autoencoders, can be used for data compression. By learning to represent high-dimensional data in a lower-dimensional latent space, autoencoders can compress data, making it easier to process and store.
- **Example:** For LHC data, autoencoders can be used to compress raw detector data into more manageable forms while still retaining key features that are necessary for analysis. This reduced representation can then be used in downstream machine learning models or analysis

pipelines.

B. Types of Neural Networks Used in Particle Physics

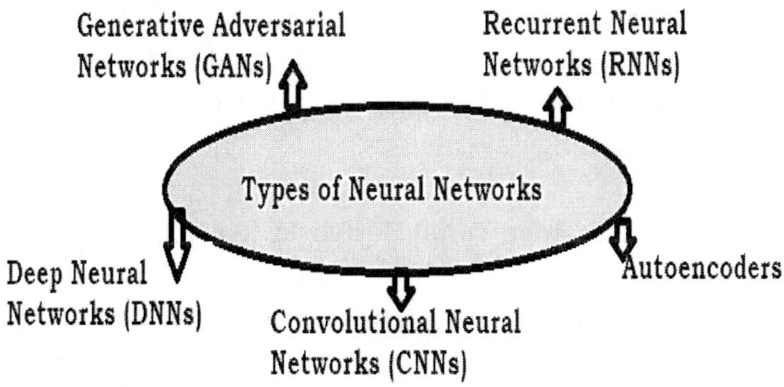

- **Deep Neural Networks (DNNs)**

Deep neural networks are powerful tools for supervised learning tasks in particle physics, such as event classification and regression. By using multiple hidden layers, DNNs can model complex, nonlinear relationships in high-dimensional data. DNNs are commonly used for tasks like signal classification, energy regression, and other predictive tasks.

- **Convolutional Neural Networks (CNNs)**

CNNs are particularly effective for image and grid-like data, which makes them ideal for processing data from particle detectors, where information is often represented in the form of grids or images (e.g., calorimeter hits). CNNs excel in **feature extraction** and **pattern recognition**, making them particularly useful in tasks like event classification and object detection in particle collision data.

- **Recurrent Neural Networks (RNNs)**

RNNs are useful for time-series data, where the order of events matters. While RNNs are not as commonly used in particle physics, they can be applied to analyze temporal data or sequences of events, such as the progression of particle interactions over time.

- **Autoencoders**

Autoencoders are unsupervised learning models that aim to learn a compact representation of input data through an encoding-decoding process. They are particularly useful for anomaly detection, where the model learns to reconstruct normal data and can flag deviations that may indicate interesting new physics.

- **Generative Adversarial Networks (GANs)**

GANs consist of two neural networks: a generator and a discriminator, that are trained together in a process of adversarial competition. In particle physics, GANs can be used to generate realistic simulations of particle interactions or detector data, helping to supplement the experimental data or accelerate the simulation process.

3.4 Machine Learning in Astrophysics

Astrophysics, the branch of astronomy that studies the physical properties and interactions of celestial objects and phenomena, has been significantly transformed by the application of **machine learning (ML)** techniques. The vast amounts of data generated by modern astrophysical surveys, space telescopes, and particle detectors make it difficult for traditional methods to efficiently handle and extract meaningful insights. Machine learning, with its ability to handle large datasets, identify patterns, and make predictions, has become an essential tool in the study of the universe.

A. Key Applications of Machine Learning in Astrophysics

1. Galaxy and Star Classification

One of the most important tasks in astrophysics is the classification of galaxies and stars based on their physical properties, such as size, luminosity, and spectral features. Traditionally, this was done through manual inspection of images and spectra, but with modern datasets, this is no longer feasible.

- **Machine Learning in Galaxy Classification:** ML algorithms, such as **support vector machines (SVMs)**, **decision trees**, and **neural networks**, have been employed to classify galaxies based on their morphology (e.g., spiral, elliptical). Deep learning techniques, such as **convolutional neural networks (CNNs)**, have shown excellent performance in classifying galaxies from large image datasets, such as those from the **SDSS (Sloan Digital Sky Survey)** or **Hubble Space Telescope**.
- **Example:** The **Galaxy Zoo** project is a citizen science initiative where volunteers classify galaxies. ML has been used to automate and improve this process by training algorithms to match human-classified galaxies and assist in distinguishing rare galaxy types or those with complex morphologies.

2. Supernova Detection and Classification

Supernovae are explosive events marking the death of certain types of stars. Their study provides critical information about the life cycles of stars, the expansion of the universe, and the properties of dark energy.

- **Automated Supernova Detection:** With the rise of sky surveys like **Pan-STARRS**, **Zwicky Transient Facility (ZTF)**, and **LSST (Large Synoptic Survey Telescope)**, millions of images are produced regularly. Machine learning is used to identify transient objects (such as supernovae) within these images by detecting unusual light curves or patterns of brightness change.
- **Example:** Deep learning models can classify supernovae based on their light curves, distinguishing between different types (e.g., Type Ia, Type II). This is crucial for understanding the different explosion mechanisms and the role of supernovae in stellar evolution.

3. Gravitational Wave Detection

Gravitational waves are ripples in spacetime caused by massive objects like merging black holes or neutron stars. The detection of gravitational waves opened a new observational window into the universe. The challenge lies in separating the weak gravitational wave signals from noise in the data.

- **Machine Learning for Signal Detection:** ML is used to detect gravitational waves by analyzing data from observatories like **LIGO** and **Virgo**. Machine learning algorithms, such as **neural networks** and **random forests**, are trained to distinguish between genuine gravitational wave signals and background noise.

- **Example:** In the first detection of gravitational waves (from the merger of two black holes), machine learning algorithms helped in filtering out noise and identifying the true signal. Ongoing efforts to detect more events involve the use of ML for real-time data analysis.

4. Cosmic Microwave Background (CMB) Analysis

The **cosmic microwave background (CMB)** is the afterglow of the Big Bang and provides critical insights into the early universe. However, the analysis of CMB data is complicated by various sources of noise and systematic errors.

- **ML for CMB Anomaly Detection:** Machine learning is applied to identify anomalies in the CMB data, model the statistical properties of the background, and extract features such as **baryon acoustic oscillations** or **reionization epoch** that can shed light on cosmological parameters.
- **Example:** The **Planck satellite** provided CMB data, and ML techniques were employed to separate noise from real cosmological signals. **Convolutional neural networks (CNNs)** are being tested to better understand the data and improve signal-to-noise ratios.

5. Exoplanet Detection

The discovery of exoplanets (planets orbiting stars outside our solar system) is one of the most exciting areas of modern astrophysics. The most successful method for detecting exoplanets is the **transit method**, where the dimming of a star's light is measured as a planet passes in front of it.

- **Machine Learning in Exoplanet Detection:** Machine learning algorithms are widely used to process data from telescopes like **Kepler** and **TESS**. ML helps identify periodic dips in brightness that correspond to exoplanet transits, and it is also used for classifying exoplanet candidates based on their light curves.

• **Example:** The **Kepler Space Telescope** generated a large dataset of star light curves, and ML algorithms were used to identify potential exoplanet transits. **Deep neural networks (DNNs)** and **decision trees** have been particularly useful in making predictions about the likelihood of a given candidate being an actual exoplanet.

6. Dark Matter and Dark Energy Research

One of the most significant mysteries in astrophysics is the nature of **dark matter** and **dark energy**. Machine learning is helping scientists study these phenomena by analyzing data from galaxy surveys, gravitational lensing, and large-scale structure measurements.

• **ML in Dark Matter Studies:** ML models, such as **random forests** and **neural networks**, are used to identify subtle correlations in large-scale structure data, gravitational lensing patterns, and galaxy clustering that could provide clues about the presence and distribution of dark matter.

• **Example:** In **weak gravitational lensing**, the light from distant galaxies is distorted by the gravitational field of intervening dark matter. ML is used to analyze these distortions and map the dark matter distribution across the universe.

B. Machine Learning Techniques Used in Astrophysics

- **Supervised Learning**

Supervised learning involves training a model on labeled data to make predictions on unseen data. In astrophysics, supervised learning is used for tasks like **classification** (e.g., galaxy types, supernova types) and **regression** (e.g., determining the redshift of galaxies from spectral data).

Example: Classifying stars based on their spectra or determining the age of galaxies from observed features.

- **Unsupervised Learning**

Unsupervised learning methods do not require labeled data. These methods are useful for exploring complex datasets, identifying patterns, or clustering similar objects.

Example: Clustering galaxies based on their physical properties or finding unusual patterns in large datasets, such as anomalies in the cosmic microwave background or galaxy distributions.

- **Deep Learning**

Deep learning, a subset of machine learning based on neural networks with many layers, is particularly effective for handling complex, high-dimensional datasets such as images or time-series data.

Example: Convolutional Neural Networks (CNNs) for galaxy and supernova classification or using **Recurrent Neural Networks (RNNs)** for analyzing time-series data from variable stars or light curves.

- **Reinforcement Learning**

Reinforcement learning (RL) involves training algorithms to make decisions by interacting with an environment. In astrophysics, RL can be used for tasks like optimizing survey strategies or analyzing telescope observations.

Example: Optimizing the scheduling of telescope observations to maximize the discovery of rare astronomical events.

C. Challenges of Machine Learning in Astrophysics

- **Data Volume and Quality:** Modern astrophysical surveys generate enormous amounts of data, requiring machine learning algorithms to be both computationally efficient and accurate. Additionally, astronomical data can be noisy, requiring robust models that can handle imperfections in the data.
- **Interpretability:** Many ML algorithms, especially deep learning models, are often considered "black boxes," meaning their decision-making process is not easily interpretable. In astrophysics, understanding why a model made a particular prediction is crucial for confirming scientific hypotheses.
- **Generalization:** Machine learning models trained on one dataset (e.g., data from one telescope) may not generalize well to other datasets (e.g., from a different instrument or in a different part of the sky). Ensuring that models can handle a wide range of observational conditions is a key challenge.
- **Small Sample Sizes:** Some astrophysical phenomena, such as rare events (e.g., gamma-ray bursts), may have very few labeled samples. In such cases, supervised learning is difficult, and techniques like transfer

learning or semi-supervised learning are often used.

D. The Future of Machine Learning in Astrophysics

As machine learning continues to evolve, its role in astrophysics is expected to grow exponentially. Some potential future developments include:

- **Real-time Data Analysis:** Machine learning will play a crucial role in processing data in real-time, especially from telescopes and space observatories that generate continuous streams of data.
- **Quantum Machine Learning:** With the advent of quantum computing, quantum machine learning techniques could accelerate astrophysical simulations and data analysis tasks, potentially solving complex problems faster than classical computers.
- **Multi-Modal Data Integration:** The combination of various types of data (e.g., gravitational waves, optical images, and radio waves) will require machine learning models that can integrate and make sense of data from multiple sources to form a more complete picture of astronomical phenomena.

Machine learning has become a transformative tool in **astrophysics**, enabling researchers to efficiently analyze vast datasets, discover new phenomena, and gain deeper insights into the workings of the universe. As the field continues to develop, the synergy between machine learning and astrophysical research promises to unlock many more secrets of the cosmos, from the nature of dark matter to the exploration of distant exoplanets.

3.5 Chaos Theory and Nonlinear Dynamics

Chaos theory and nonlinear dynamics are branches of physics that study systems that are highly sensitive to initial conditions, leading to behavior that appears random but is actually deterministic. These systems, although governed by deterministic laws, exhibit irregular, unpredictable, and complex behaviors. Traditional methods of analysis often struggle with these systems due to their sensitivity to initial conditions and the complexity of the underlying equations.

Machine learning (ML), with its ability to uncover patterns in large and complex datasets, has shown great promise in advancing the study of chaos and nonlinear dynamics. In particular, ML techniques are useful for analyzing chaotic systems, forecasting their future states, identifying patterns, and even controlling chaotic behavior.

Here's an overview of how machine learning is being integrated into chaos theory and nonlinear dynamics in physics.

A. Understanding Chaos and Nonlinear Dynamics

- **Chaos Theory**

Chaos theory focuses on the behavior of dynamic systems that are **deterministic** but exhibit **sensitive dependence on initial conditions**, which means small changes in the initial state of the system can lead to vastly different outcomes. This property is commonly known as the **butterfly effect**. Chaotic systems often show unpredictable and random-like behavior over time.

- **Examples of chaotic systems:** Weather systems, fluid dynamics, population dynamics in ecology, and the double pendulum are classic examples of chaotic systems in physics.

- **Nonlinear Dynamics**

Nonlinear dynamics deals with systems whose behavior cannot be described by linear equations. Unlike linear systems, where the principle of superposition holds, nonlinear systems exhibit complex behaviors like bifurcations, limit cycles, and chaotic attractors.

- **Examples of nonlinear systems:** Pendulums (when driven at the right frequency), neural networks, chemical reactions (e.g., the Belousov-Zhabotinsky reaction), and economic systems.

B. Role of Machine Learning in Chaos and Nonlinear Dynamics

Machine learning algorithms have proven effective for studying chaotic and nonlinear systems in several key areas, including modeling, prediction, pattern recognition, control, and optimization.

1. Identifying Patterns in Chaos

One of the challenges of studying chaotic systems is the difficulty in predicting future states because of their sensitive dependence on initial conditions. However, chaos often shows hidden patterns and structures within seemingly random behavior.

• **Example:** Machine learning models, such as **recurrent neural networks (RNNs)** and **long short-term memory networks (LSTMs)**, are particularly useful for time-series prediction. These models are capable of learning patterns in chaotic systems by analyzing the system's historical behavior.

Application: Forecasting chaotic time series data such as weather patterns, stock market prices, or fluid dynamics simulations. ML models can extract underlying trends that are otherwise difficult to model with traditional techniques.

2. Model Discovery and System Identification

In nonlinear dynamics, it is often difficult to derive the governing equations of a system, especially when only noisy or incomplete data is available. ML, specifically **symbolic regression** and **neural networks**, can be used to **discover underlying models** directly from data.

• **Example: Neural networks** can be trained to approximate the behavior of chaotic systems, even when the precise physical model is unknown. **Symbolic regression** algorithms (e.g., **genetic programming**) can be used to search for mathematical expressions that describe the system's dynamics based on input-output data.

• **Application:** ML has been used in systems like coupled oscillators or fluid turbulence to approximate the governing nonlinear differential equations without having explicit models for them.

3. Time Series Prediction and Chaos Forecasting

One of the most important applications of machine learning in chaos theory is **predicting future states of chaotic systems**. The key challenge here is that, due to the sensitive dependence on initial conditions, predicting future behavior with high accuracy is difficult.

• **Example: LSTM networks**, which are designed to handle sequential data, are effective in making time-step predictions for chaotic systems, such as the prediction of atmospheric conditions, population dynamics, or financial time series.

- **Application:** In weather forecasting, LSTM networks have been used to predict future weather patterns based on previous data, a problem that traditional forecasting methods struggle with due to the nonlinear nature of the climate system.

4. Control of Chaos

Chaotic systems, despite their unpredictability, can often be controlled or stabilized using nonlinear control strategies. ML techniques can be employed to design adaptive control methods that steer chaotic systems toward a desired state or suppress unwanted chaotic behavior.

- **Example: Reinforcement learning (RL)**, a type of machine learning in which an agent learns to make decisions by interacting with its environment, has been applied to control chaotic systems. RL algorithms can optimize control strategies to suppress chaos or stabilize unstable periodic orbits (also known as **chaotic attractors**).
- **Application:** In physical systems such as lasers, electrical circuits, and chemical reactors, RL has been used to stabilize chaotic oscillations or to guide the system toward periodic behavior.

5. Predicting Bifurcations and Phase Transitions

Bifurcations are critical points in nonlinear systems where a small change in parameters causes a dramatic change in the system's behavior. Machine learning can be used to detect bifurcations or predict **phase transitions** in complex systems.

- **Example: Deep learning algorithms** can be trained to predict when a system will undergo a bifurcation based on parameter space exploration. ML can analyze high-dimensional datasets to detect subtle changes in system dynamics, enabling early detection of bifurcations that might be difficult to identify using traditional methods.
- **Application:** ML techniques have been used to predict phase transitions in systems like magnetism, fluid dynamics, and even in the study of the brain's neural activity.

6. Optimization of Nonlinear Systems

Optimization is a crucial aspect of understanding and controlling nonlinear systems. Machine learning techniques such as **genetic algorithms (GAs)**, **particle swarm optimization (PSO)**, and **neural networks** are applied to optimize complex nonlinear systems.

Example: Genetic algorithms (GA), which mimic the process of natural evolution, are particularly effective at finding optimal solutions in systems with highly complex, nonlinear objective functions.

Application: In fluid dynamics, GAs have been used to optimize the design of flow systems. In nonlinear chemical reactions, ML algorithms can be used to find conditions that maximize reaction rates or efficiency.

C. ML Algorithms Applied to Chaos and Nonlinear Systems

1. Neural Networks

- **Feedforward Neural Networks (FNNs):** Used for system identification and approximation of nonlinear dynamical systems.
- **Recurrent Neural Networks (RNNs) and LSTMs:** Effective in learning long-term dependencies and making predictions in time series data from chaotic systems.

2. Reinforcement Learning (RL)

- RL algorithms have been used to stabilize chaotic systems and optimize their behavior, particularly in controlling systems with unknown or complex dynamics.

3. Support Vector Machines (SVMs)

- SVMs can be applied for classification tasks in chaotic systems, such as detecting bifurcations or classifying different types of chaotic behaviors.

4. Symbolic Regression and Genetic Programming

- These techniques are used to automatically discover the governing equations of chaotic and nonlinear systems from data.

5. Cluster Analysis

- **K-means** and **DBSCAN** clustering algorithms can be applied to detect regimes in chaotic systems, identifying periods of regular vs. chaotic behavior or identifying different attractor states.

D. Challenges and Future Directions

1. Interpretability

One of the key challenges of using ML in chaotic systems is the lack of interpretability of models, especially with deep learning techniques. Physicists often require insight into the underlying mechanisms and dynamics of chaotic behavior, and ML models can be difficult to interpret.

2. Overfitting

Chaos and nonlinear systems often involve high-dimensional, noisy data, making them prone to overfitting. Ensuring that machine learning models generalize well to new, unseen data is a key challenge in predicting chaotic systems.

3. Data Quality

Chaotic systems can produce noisy data that is sensitive to initial conditions, and the quality of the data can significantly impact the performance of machine learning models. This requires robust techniques to handle noise and outliers in the data.

4. Scaling

As chaotic systems become more complex or as the number of dimensions increases, the computational cost of training machine learning models also increases. Finding scalable techniques that can efficiently handle high-dimensional data from nonlinear systems is an ongoing challenge.

Machine learning offers powerful tools for studying chaos and nonlinear dynamics, which are fundamental to understanding many physical systems. ML algorithms help physicists model, predict, and control chaotic behavior, uncover hidden patterns, and optimize nonlinear systems. As ML techniques continue to evolve, their application to chaos theory and nonlinear dynamics is expected to deepen, leading to new discoveries and advancements in understanding complex, dynamical systems.

3.6 Machine Learning in Thermodynamics and Statistical Mechanics

A. Machine Learning in Thermodynamics

Thermodynamics, the branch of physics that deals with the relationships between heat, work, temperature, and energy, traditionally relies on well-established laws and equations such as the laws of thermodynamics, the ideal gas law, and various statistical mechanics models. These principles have enabled scientists to understand and predict the behavior of macroscopic systems. However, many complex thermodynamic systems, especially those involving large numbers of interacting particles, are computationally intractable using traditional methods.

Machine learning (ML) has emerged as a powerful tool to address these challenges. By analyzing large datasets and identifying hidden patterns, ML techniques can provide valuable insights into complex thermodynamic systems, help discover new materials, optimize energy processes, and even aid in solving problems that were once deemed too difficult for conventional computational methods.

B. Applications of Machine Learning in Thermodynamics

- **Predicting Thermodynamic Properties** One of the most common applications of machine learning in thermodynamics is the prediction of thermodynamic properties such as pressure, temperature, entropy, and enthalpy for complex systems.
- **Machine Learning for Material Properties:** Traditional methods, like Monte Carlo simulations or molecular dynamics, are often computationally expensive, especially when studying large systems. ML models, such as **neural networks** and **support vector machines (SVMs)**, can be trained on existing datasets of thermodynamic properties to predict the properties of new materials or systems.
- **Example:** ML algorithms can predict the **thermal conductivity**,

heat capacity, or **viscosity** of new materials by learning from experimental or simulated data. This approach has been used to design advanced materials for energy storage or thermal management.

- **Phase Transitions and Critical Phenomena**

Phase transitions (e.g., from liquid to gas) are a central topic in thermodynamics. These transitions often involve critical points where small changes in variables like pressure or temperature lead to large-scale changes in a system's behavior. Traditional methods to study phase transitions can be very complex, especially near the critical point.

Machine Learning for Phase Transition Detection: Machine learning can help identify and characterize phase transitions in complex systems. For example, **unsupervised learning** algorithms like **autoencoders** or **principal component analysis (PCA)** can detect critical behavior in high-dimensional thermodynamic data and classify different phases (e.g., gas, liquid, or solid).

Example: Deep neural networks (DNNs) have been used to study critical phenomena and to predict phase transitions in systems such as spin glasses, ferromagnetic systems, and fluid dynamics. These models can recognize patterns in systems that are hard to detect with traditional techniques.

- **Optimizing Energy Conversion and Efficiency**

In thermodynamics, energy conversion and efficiency are key concerns, particularly in applications like engines, refrigeration systems, and renewable energy. ML can assist in optimizing these processes to achieve higher efficiency and performance.

- **Machine Learning for Thermodynamic Systems Optimization:** Reinforcement learning (RL), a type of machine learning where agents learn to make decisions by interacting with an environment, has been used to optimize the design and operation of thermodynamic systems. For example, RL can be used to optimize the operation of heat engines, refrigeration cycles, or chemical reactors, thereby improving their energy efficiency.
- **Example:** In **heat engines**, ML algorithms can be applied to

optimize the thermodynamic cycle by adjusting operating parameters to maximize work output or minimize fuel consumption. Similarly, RL can optimize **thermochemical processes** in industrial settings, such as material production and chemical synthesis.

- **Entropy and Information Theory**

Entropy, a central concept in thermodynamics, measures the disorder or randomness of a system. In statistical mechanics, entropy is connected to the number of possible microstates of a system. Recently, the concept of **information entropy** has been explored through the lens of machine learning.

- **Machine Learning for Entropy Calculation:** ML techniques, particularly **neural networks** and **unsupervised learning**, have been applied to estimate the entropy of complex systems. For example, ML models can be used to calculate **Shannon entropy** in data-driven models, improving our understanding of information entropy in thermodynamic systems.
- **Example:** In **complex fluid systems**, ML models can estimate the entropy change during phase transitions, allowing for better control and prediction of material properties.

- **Thermodynamic Integration and Simulation**

Thermodynamic integration is a computational technique used to calculate properties such as free energy differences between different states of a system. This process can be computationally expensive, especially for systems with many interacting particles.

- **Machine Learning for Simulation and Integration:** Machine learning methods such as **Gaussian processes (GPs)** and **deep learning models** can enhance traditional methods of thermodynamic integration by predicting the free energy changes and speeding up simulations. These models can provide approximations to free energies without requiring exhaustive simulations of particle interactions.
- **Example:** ML algorithms have been used to reduce the computational cost of calculating free energy differences in molecular simulations, allowing for more efficient analysis of large systems like

proteins or polymers.

C. Machine Learning Techniques Used in Thermodynamics

1. Supervised Learning

Supervised learning is often applied in thermodynamics when labeled data is available. For instance, models such as **regression models** (e.g., support vector regression) and **neural networks** can be trained to predict thermodynamic properties from known data.

- **Example:** Predicting the **heat capacity** of a material based on its temperature and pressure using supervised learning models.

2. Unsupervised Learning

Unsupervised learning techniques are useful for identifying patterns and structures in data without prior labels. These methods are particularly helpful for discovering new phase transitions, clustering systems, or detecting new thermodynamic states.

- **Example: Principal component analysis (PCA)** or **autoencoders** can be used to reduce the dimensionality of complex thermodynamic data, enabling more efficient analysis of systems with multiple variables.

3. Reinforcement Learning (RL)

Reinforcement learning, where algorithms learn by trial and error through interaction with their environment, is well-suited for optimizing thermodynamic systems such as engines or refrigerators.

- **Example:** Optimizing the operation of a **Carnot engine** or a **refrigeration cycle** by adjusting input parameters (such as temperature or pressure) to maximize efficiency or minimize energy consumption.

4. Neural Networks and Deep Learning

Deep learning models, particularly **convolutional neural networks (CNNs)** and **recurrent neural networks (RNNs)**, are used to handle complex datasets and make predictions in thermodynamics. These models are effective in pattern recognition, time-series prediction, and feature extraction from high-dimensional data.

- **Example: Deep neural networks** can predict thermodynamic quantities such as phase boundaries or entropy changes for complex systems like liquid crystals or polymers.

5. Gaussian Processes (GPs)

Gaussian processes are used in thermodynamics to model and predict continuous variables, such as free energy or phase transitions, with uncertainty quantification.

- **Example:** Using GPs to predict the **free energy differences** between different states of a thermodynamic system, providing a more accurate and faster method than traditional Monte Carlo or molecular dynamics simulations.

D. Challenges of Machine Learning in Thermodynamics

While machine learning has great potential in thermodynamics, there are several challenges that need to be addressed:

- **Data Quality and Availability**

Machine learning algorithms require large, high-quality datasets to perform well. In thermodynamics, especially when studying new materials or phenomena, obtaining reliable and comprehensive datasets can be difficult.

- **Model Interpretability**

ML models, particularly deep learning techniques, are often considered "black boxes," meaning that their decision-making processes are not easily interpretable. In thermodynamics, where understanding the underlying physical mechanisms is important, the lack of interpretability can be a barrier.

- **Computational Complexity**

Thermodynamic simulations, especially at the molecular level, are computationally expensive. ML models, while they can accelerate predictions, may still require large amounts of computational resources to train, especially for systems with many variables or complex dynamics.

- **Generalization**

ML models trained on specific thermodynamic systems may struggle to generalize to new, unseen systems. Developing robust models that can perform well across different types of systems remains an open challenge.

As machine learning techniques continue to evolve, their integration with thermodynamics holds great potential:

- **Improved Materials Discovery:** ML will continue to play a significant role in predicting new materials with specific thermodynamic properties, such as high thermal conductivity, low thermal expansion, or efficient heat storage.
- **Energy Optimization:** Machine learning will help optimize energy conversion processes, reduce waste, and design systems that operate closer to their theoretical efficiency limits.
- **Integrated Modeling and Simulation:** ML will assist in creating more efficient simulation models for complex thermodynamic systems, including multiscale systems and systems with many degrees of freedom.
- **Interdisciplinary Approaches:** As ML continues to merge with thermodynamics, interdisciplinary approaches that combine physics, computer science, and engineering will lead to innovative solutions in energy systems, material science, and complex system analysis.

Machine learning is revolutionizing the field of thermodynamics by providing tools to predict thermodynamic properties, optimize energy systems, detect phase transitions, and improve material discovery. While challenges remain, the integration of ML into thermodynamics is opening up new possibilities for solving complex problems, designing efficient systems, and understanding the fundamental principles governing energy and matter. As ML techniques continue to develop, their potential in thermodynamics will only continue to grow, driving innovation in energy, materials, and complex system analysis.

D. Machine Learning in Statistical Mechanics

Statistical Mechanics is a branch of physics that uses statistical methods to explain and predict the behavior of systems composed of a large number of particles. It bridges the microscopic world of individual particles and the macroscopic observable quantities such as temperature, pressure, and volume. Central to statistical mechanics are concepts like **entropy**, **free energy**, **partition functions**, and **probability**

distributions that describe the likelihood of different states of a system.

In recent years, **machine learning (ML)** has become an increasingly powerful tool for solving problems in statistical mechanics. By leveraging data-driven techniques, ML has provided new ways to analyze complex systems, uncover hidden patterns, and predict system behavior, often outperforming traditional analytical methods or enhancing their computational efficiency.

This article explores how machine learning is being used in statistical mechanics, with a focus on applications, methods, and the challenges that come with integrating ML into this field.

1. Applications of Machine Learning in Statistical Mechanics

- **Learning and Predicting Thermodynamic Properties**

Statistical mechanics aims to derive thermodynamic properties (e.g., entropy, temperature, free energy) from the microscopic states of a system. Machine learning can assist by providing alternative, data-driven methods to approximate these quantities.

- **Example: Neural networks** and other ML models can be trained to predict thermodynamic properties from the configuration of particles in a system, such as predicting **free energy**, **heat capacity**, and **equation of state** for materials that have complex interactions and behavior.

- **Phase Transitions and Critical Phenomena**

Phase transitions, such as the change from a liquid to a gas, are central to statistical mechanics. In many systems, these transitions are accompanied by critical phenomena—sharp changes in behavior or fluctuations near the transition point.

- **Machine learning for phase transition detection:** ML techniques can be used to detect and characterize phase transitions in systems that are difficult to analyze using traditional methods. For example, **deep neural networks** (DNNs) and **unsupervised learning** techniques like **autoencoders** or **principal component analysis (PCA)** can uncover hidden critical points in the data, identifying the transition points that separate different phases of matter.
- **Example:** ML has been applied to study the **Ising model** and other

spin systems to detect critical temperatures and phase boundaries. ML models can help predict these transitions without relying on traditional scaling arguments or direct numerical simulations.

a)Dynamics of Complex Systems

Statistical mechanics deals with the statistical properties of systems of many interacting particles. These systems often exhibit complex behavior, including long-range correlations, non-equilibrium dynamics, and memory effects.

- **Machine learning for time-series prediction and modeling:** ML algorithms, such as **recurrent neural networks (RNNs)** and **long short-term memory networks (LSTMs)**, can be used to model the temporal evolution of these systems. These models are capable of learning long-range dependencies, helping predict future states of the system based on past behavior.
- **Example:** For systems like **fluids** or **plasmas**, ML can be used to predict the evolution of macroscopic properties such as temperature or pressure over time.
- **Accelerating Monte Carlo and Molecular Dynamics Simulations**

Monte Carlo (MC) and molecular dynamics (MD) simulations are fundamental tools in statistical mechanics, used to sample the states of a system and compute macroscopic properties. However, these simulations can be computationally expensive, especially for large systems with many degrees of freedom.

- **Machine learning for simulation acceleration:** ML techniques such as **neural networks** or **Gaussian processes** can accelerate MC or MD simulations by predicting system behavior based on prior simulation data. ML models can be trained to replace or guide the simulation process, reducing the number of steps required to converge to an equilibrium state.
- **Example:** ML methods like **reinforcement learning (RL)** can guide Monte Carlo simulations by exploring the most relevant parts of the configuration space more efficiently, reducing the time required to calculate macroscopic quantities.

- **Discovery of Novel Materials and States of Matter**

Statistical mechanics plays a crucial role in the discovery and design of new materials. Machine learning can be used to mine large databases of materials and predict their thermodynamic properties or other key characteristics based on their microscopic structure.

- **Machine learning for materials design:** ML can be applied to find new materials with specific properties by predicting their behavior based on microscopic characteristics, such as atomic structure and bonding. For instance, **graph neural networks (GNNs)** are particularly effective at predicting material properties by capturing the local and global connectivity between atoms in a material.
- **Example:** ML is used to predict **phase diagrams** of materials, find new superconductors, or design alloys with desired properties by learning from the data of known compounds.

E. Machine Learning Techniques in Statistical Mechanics

- **Supervised Learning**

In supervised learning, the model is trained on labeled data, where both the input (system configuration) and output (thermodynamic properties) are known. This technique can be used to predict macroscopic properties such as free energy or temperature from microscopic data.

- **Example:** Using **support vector machines (SVMs)** or **neural networks** to predict thermodynamic properties like **entropy**, **enthalpy**, and **pressure** from molecular configurations or experimental data.

- **Unsupervised Learning**

Unsupervised learning methods are valuable when dealing with large datasets where the relationships between variables are not fully known. These methods can be used to identify hidden patterns, clusters, or structures in the data.

- **Example: Principal component analysis (PCA)** or **autoencoders** can be used to reduce the dimensionality of thermodynamic data and uncover latent variables that are important for understanding phase transitions or critical points in complex systems.

- **Reinforcement Learning (RL)**

Reinforcement learning is a branch of machine learning where an agent learns to make decisions by interacting with an environment. In statistical mechanics, RL can be used to optimize simulation processes, accelerate sampling, or discover efficient pathways in energy landscapes.

- **Example: Q-learning** and **actor-critic methods** can guide Monte Carlo simulations or molecular dynamics simulations toward more efficient exploration of state space, reducing the computational cost of sampling thermodynamic quantities.

- **Deep Learning**

Deep learning, particularly **deep neural networks (DNNs)**, has become a powerful tool for learning complex mappings from microscopic to macroscopic quantities. DNNs are particularly useful in modeling highly non-linear relationships and extracting features from large datasets.

- **Example: Convolutional neural networks (CNNs)** have been used to analyze spatial structures of materials, while **LSTMs** are particularly suited for time-series prediction of dynamical systems, helping model and predict the behavior of systems far from equilibrium.

- **Gaussian Processes (GPs)**

Gaussian processes are a type of probabilistic model used for regression and classification tasks, where predictions are associated with uncertainty estimates. They are particularly useful in scenarios where the function being modeled is unknown but can be inferred from data.

- **Example:** GPs are used to estimate thermodynamic quantities, such as free energy differences, from a set of known data, and they provide uncertainty estimates, which can guide further simulations or experiments.

F. Challenges and Limitations

1. Data Availability and Quality

Machine learning models require large amounts of high-quality data for training, but obtaining reliable and representative datasets for complex systems can be challenging. Data from simulations may be noisy, incomplete, or unbalanced, which can negatively affect the model's performance.

2. Interpretability

While machine learning models, particularly deep neural networks, can make highly accurate predictions, they are often considered "black-box" models. In physics, it is important not only to make predictions but also to understand the underlying physical processes. This is a significant challenge when applying ML to statistical mechanics, as many ML models provide little insight into the underlying physical mechanisms.

3. Generalization

ML models are often trained on specific datasets, which may limit their ability to generalize to new, unseen systems. In statistical mechanics, where systems can vary widely, it is important for ML models to be flexible enough to handle diverse scenarios while maintaining accuracy.

4. Computation Cost

Training machine learning models, particularly deep learning models, can be computationally expensive and time-consuming. For large systems, the number of particles and interactions can make it impractical to apply certain ML techniques without access to substantial computational resources.

The integration of machine learning and statistical mechanics is still in its early stages, but several promising directions are emerging:

- **Explainable AI (XAI):** Developing techniques to improve the interpretability of machine learning models could help scientists understand the physical principles behind complex systems while making accurate predictions.
- **Multiscale Modeling:** ML can help bridge different scales of modeling, from microscopic particle interactions to macroscopic thermodynamic quantities, creating more unified models for complex systems.
- **Hybrid Models:** Combining traditional statistical mechanics methods with machine learning models can lead to more efficient and

accurate predictions, especially for systems with complex, unknown dynamics.

- **Data-Driven Discoveries:** ML has the potential to uncover new physical phenomena or identify previously unknown relationships in thermodynamic systems by analyzing large datasets from experiments or simulations.

Machine learning is rapidly becoming a valuable tool in statistical mechanics, offering new ways to predict thermodynamic properties, optimize simulations, and study complex systems. By leveraging the power of data-driven methods, ML can enhance traditional approaches to understanding phase transitions, critical phenomena, and other complex behaviors. Although challenges such as data quality, interpretability, and computational cost remain, the integration of ML in statistical mechanics is expected to lead to significant advancements in material science, thermodynamics, and complex system modeling in the coming years.

3.7 Image Processing in Experimental Physics

In **experimental physics**, image processing is a critical tool for analyzing, interpreting, and extracting meaningful data from visual information collected during experiments. Whether working with microscopes, particle detectors, or astronomical telescopes, the images obtained are often complex, noisy, and require various techniques to enhance, filter, and extract relevant features. Image processing techniques help convert raw image data into quantitative results, making it easier for physicists to test hypotheses, validate models, and make discoveries.

A. Applications of Image Processing in Experimental Physics

- **Particle Physics**

In particle physics experiments, large detectors such as those used in particle colliders (e.g., the **Large Hadron Collider** at CERN) generate vast amounts of data in the form of images. These detectors track particle interactions, decay products, and their trajectories.

- **Tracking and Reconstruction:** Image processing is used to reconstruct the paths of particles from the data collected by detectors, such as **track detectors** or **calorimeters**. This involves detecting the shapes, angles, and positions of particle tracks in a detector's visual field.

- **Example:** In experiments like **ATLAS** or **CMS** at CERN, machine learning algorithms and image processing techniques are used to identify particle tracks, measure their energy, and reconstruct the momentum of particles based on the images captured by detectors.

- **Optical Microscopy in Condensed Matter Physics**

In condensed matter physics, optical microscopy techniques like **fluorescence microscopy** and **scanning electron microscopy (SEM)** are used to study materials at the microscopic level. These techniques generate detailed images of materials, where the features can vary on a nanoscale.

- **Image Segmentation:** Image processing algorithms are applied to segment regions of interest within the images to study material properties, surface morphology, and defects in materials at the nanoscale.
- **Example:** Analyzing images of crystalline materials obtained through SEM to identify defects, grain boundaries, and other structural characteristics that may affect material properties, such as conductivity, strength, or elasticity.

- **Astronomy and Astrophysics**

In **astronomy**, telescopes capture images of distant celestial objects, including galaxies, stars, nebulae, and other cosmic phenomena. These images often contain vast amounts of noise and require advanced processing to identify and measure objects of interest.

- **Noise Reduction and Enhancement:** Image processing techniques such as **Gaussian blurring**, **median filtering**, and **contrast enhancement** are used to improve the quality of images by reducing noise and enhancing the features of interest.
- **Example:** In **Hubble Space Telescope** images, image processing techniques are used to enhance the resolution, reduce light pollution, and emphasize features such as star clusters or galaxy shapes, helping astrophysicists study the structure and behavior of distant objects.

- **Tomography in Medical Physics**

In **medical physics**, **X-ray tomography** (CT scans) and **positron emission tomography (PET)** are widely used for imaging the internal structures of the human body. These images are then processed to extract quantitative information for diagnosis and treatment planning.

- **Reconstruction and Visualization:** Image processing algorithms reconstruct 3D models from the slice images acquired by tomography. This is essential in analyzing the internal structure of tissues, organs, and bones.
- **Example:** In cancer detection, image processing algorithms are used to process CT or MRI scans to identify and track the growth of tumors, helping radiologists and oncologists in diagnosis and treatment planning.

- **Surface Profiling and Metrology**

In experimental physics, the analysis of surface features is crucial in fields such as materials science and nanotechnology. Techniques like **atomic force microscopy (AFM)** and **scanning tunneling microscopy (STM)** generate images of the surface topography of materials at the atomic scale.

- **Surface Analysis:** Image processing techniques are used to analyze the surface roughness, texture, and topographical features of materials, which can directly influence the material's properties.

Example: Using image processing to analyze AFM images to measure the height of surface features and identify nanoscale defects that could influence the material's electrical or mechanical properties.

B. Key Image Processing Techniques Used in Experimental Physics

- **Image Preprocessing**

Preprocessing involves improving the quality of the image before applying further processing techniques. This is particularly important when dealing with noisy, low-contrast, or distorted images.

- **Noise Reduction:** Filters such as **Gaussian blur**, **median filters**, and **Wiener filtering** are commonly used to remove noise from images. This step is crucial for enhancing the accuracy of subsequent analyses.

- **Contrast Enhancement:** Techniques like **histogram equalization** and **adaptive histogram equalization (CLAHE)** are used to enhance contrast, especially when dealing with images that have low visibility or require improved differentiation of features.

- **Image Segmentation**

Image segmentation is the process of partitioning an image into multiple regions based on specific criteria, such as color, intensity, or texture. Segmentation is crucial for isolating objects or features of interest, such as particle tracks, boundaries, or defects.

- **Thresholding:** Simple thresholding techniques can segment regions of interest based on pixel intensity values. More advanced methods, such as **Otsu's method**, automatically determine the optimal threshold for separating foreground and background.
- **Edge Detection:** Algorithms like **Sobel**, **Canny**, or **Laplace operators** are used to detect boundaries between different regions in an image, aiding in feature extraction or object detection.

- **Feature Extraction and Measurement**

Once the image has been segmented into relevant regions, the next step is to extract measurable features. These could include geometric properties, such as the size, shape, and orientation of objects, or statistical properties like the distribution of pixel intensities.

- **Shape Analysis:** Techniques like **contour analysis**, **region growing**, and **Hough transform** are used to analyze the shape of objects within an image. This is critical in determining the physical characteristics of observed phenomena, such as particle tracks or material structures.
- **Quantification:** After segmenting the regions of interest, features such as **area**, **perimeter**, **centroid position**, and **aspect ratio** are extracted and used for further analysis.

- **Pattern Recognition**

Pattern recognition involves identifying specific patterns or objects within an image. Machine learning techniques, including **convolutional neural networks (CNNs)** and **support vector machines (SVMs)**, are increasingly used for object detection and classification tasks.

- **Deep Learning for Object Detection:** CNNs can be trained on labeled datasets to identify complex patterns in experimental images. This is especially useful for high-throughput experiments where large numbers of images must be processed automatically.
- **Example:** In particle physics, deep learning models can be used to detect and classify particle tracks in detector images, distinguishing between various types of particles based on their paths and energies.

- **Image Registration**

Image registration involves aligning multiple images taken from different angles, times, or modalities to create a unified view of the object or phenomenon under study.

- **Multi-view Registration:** In applications such as tomography and microscopy, image registration helps align 2D slice images into 3D reconstructions, improving the accuracy of measurements and visualizations.
- **Example:** In **MRI imaging**, different slices of the body need to be registered to build a complete 3D model of internal structures, aiding in diagnosis and planning of surgical procedures.

C. Machine Learning in Image Processing for Experimental Physics

Machine learning techniques are increasingly being used in experimental physics to automate and improve image processing tasks.

- **Deep Learning for Feature Extraction**

Deep learning models, particularly **convolutional neural networks (CNNs)**, are designed to automatically extract relevant features from images, such as particle tracks, defects in materials, or celestial bodies in astronomical images.

- **Example:** In astrophysics, CNNs have been used to detect and classify different types of galaxies, stars, and cosmic events in telescope images.

- **Object Detection and Classification**

ML models, such as **support vector machines (SVMs)** and **random**

forests, can classify objects in images based on learned features. This is useful for detecting specific phenomena in experimental images, such as particle tracks in detectors or cell structures in biological imaging.

• **Example:** In biology, machine learning has been used to identify and classify cells or tissues in microscopy images for medical research.

- **Data-Driven Image Enhancement**

ML algorithms can be trained to enhance image quality, such as removing noise, improving resolution, or even generating high-resolution images from low-resolution data, using techniques like **super-resolution**.

• **Example:** In electron microscopy, ML algorithms are employed to enhance image resolution, allowing for more detailed analysis of atomic structures.

D. Challenges and Future Directions

- **Noise and Artifacts**

Experimental images often contain noise and artifacts due to limitations in detection equipment or external factors. Dealing with these issues while preserving important features is a significant challenge.

- **Large Datasets**

With the increasing use of high-throughput imaging techniques (e.g., in biology or particle physics), the volume of image data has grown exponentially. Managing, processing, and analyzing these large datasets requires efficient algorithms and computational resources.

- **Real-Time Processing**

In some experimental setups, such as high-energy physics detectors or astronomical observations, images must be processed in real-time to trigger actions or inform decisions during the experiment. Achieving real-time performance while maintaining accuracy remains a challenge.

- **Interpretability and Automation**

While ML techniques are powerful, ensuring that models can be interpreted in a meaningful way is crucial in physics experiments, where the underlying physical processes must be understood and explained. Automating image analysis workflows, while preserving interpretability, is a key direction for future development.

Image processing plays an essential role in **experimental physics**, helping scientists extract meaningful information from complex visual data. With advances in machine learning and deep learning, the ability to process, analyze, and interpret experimental images has greatly improved. Whether it's particle detection, material analysis, or astronomical observations, image processing techniques enable physicists to gain deeper insights into the phenomena under study, drive discoveries, and improve the accuracy of experimental measurements.

Chapter 4

Machine Learning in Electronics

4.1 Automated Circuit Design and Optimization

Automated circuit design and optimization using machine learning is transforming the way electronic circuits are developed, making the process faster, more efficient, and less reliant on human expertise. Traditional circuit design is a complex task that involves selecting components, determining optimal layouts, and fine-tuning parameters to achieve desired performance characteristics. This often requires iterative testing and manual adjustments, which can be time-consuming and costly. Machine learning (ML) introduces a data-driven approach to automate and optimize circuit design, reducing development cycles and improving performance.

One of the key advantages of using ML in circuit design is its ability to explore vast design spaces and identify optimal solutions with minimal human intervention. Techniques such as deep learning, reinforcement learning, and evolutionary algorithms allow AI models to analyze existing circuit designs, predict their performance, and generate optimized configurations. For example, reinforcement learning can be used to train an AI agent to refine circuit layouts based on performance feedback, while genetic algorithms help evolve circuit designs by mimicking natural selection processes.

Bayesian optimization is another widely used technique in automated circuit design. It enables efficient exploration of design parameters by balancing exploration and exploitation, leading to faster convergence toward optimal solutions. Neural architecture search (NAS), commonly used in optimizing deep learning models, can also be applied to circuit design by automatically discovering efficient architectures tailored to specific applications.

Optimization in circuit design is crucial for meeting various performance constraints, such as power consumption, speed, and area efficiency. In applications like IoT devices, where power efficiency is critical, ML-driven optimization ensures minimal energy usage while maintaining functionality. Similarly, in high-speed computing and telecommunications, ML can optimize signal integrity and minimize noise, improving overall system performance.

Another significant benefit of ML-based circuit design is its adaptability. AI models can learn from past designs and apply those insights to new projects, reducing the need for manual adjustments. Additionally, ML techniques can assist in real-time optimization, allowing circuits to self-adjust based on environmental conditions or operational requirements.

As AI technology advances, automated circuit design and optimization will continue to revolutionize industries such as semiconductor manufacturing, automotive electronics, and telecommunications. By reducing design complexity, lowering costs, and accelerating innovation, machine learning-driven circuit design is paving the way for the next generation of electronic systems.

4.2 Fault Detection in Electronic Systems

Fault detection in electronic systems is a critical aspect of ensuring reliability, safety, and efficiency in various applications, including industrial automation, telecommunications, consumer electronics, and aerospace systems. Traditional fault detection methods rely on manual inspection, rule-based algorithms, or hardware redundancy, which can be inefficient, costly, and slow in detecting failures. Machine learning (ML) has emerged as a powerful tool to automate fault detection, improve accuracy, and reduce downtime by analyzing vast amounts of data in real time.

A. Role of Machine Learning in Fault Detection

Machine learning techniques enable automated fault detection by learning patterns from historical data and identifying deviations that indicate potential failures. The process typically involves the following steps:

1. **Data Collection** – Sensors, logs, and diagnostic tools generate real-time data from electronic systems, including voltage levels, current

flow, temperature, signal integrity, and error logs.

2. Data Preprocessing – Raw data is cleaned, normalized, and structured to remove noise and enhance feature extraction. Feature engineering techniques help identify the most relevant parameters affecting system performance.

3. Model Training – Supervised learning, unsupervised learning, and reinforcement learning models are trained using historical fault and normal operational data.

4. Fault Detection & Classification – The trained ML model continuously monitors incoming data and detects anomalies or classifies faults based on learned patterns.

5. Fault Prediction & Prevention – Advanced models, such as predictive maintenance systems, use ML to forecast potential failures before they occur, allowing for preventive actions.

B. Key Machine Learning Techniques for Fault Detection

1. Supervised Learning – Algorithms like Support Vector Machines (SVM), Decision Trees, and Neural Networks classify faults based on labeled training data.

2. Unsupervised Learning – Clustering techniques such as K-Means and Autoencoders detect anomalies by identifying outliers in unlabeled data.

3. Reinforcement Learning – Adaptive fault detection systems use reinforcement learning to improve detection efficiency over time based on reward-based learning.

4. Deep Learning – Convolutional Neural Networks (CNNs) and Recurrent Neural Networks (RNNs) are used for complex pattern recognition in time-series fault data.

C. Advantages of Machine Learning-Based Fault Detection

- **High Accuracy** – ML models continuously improve their detection accuracy through learning.
- **Real-Time Monitoring** – AI-driven systems provide real-time fault detection, reducing response time.
- **Cost-Effective Maintenance** – Predictive maintenance prevents costly repairs by detecting faults early.
- **Scalability** – ML-based fault detection systems can be easily integrated into large-scale industrial applications.

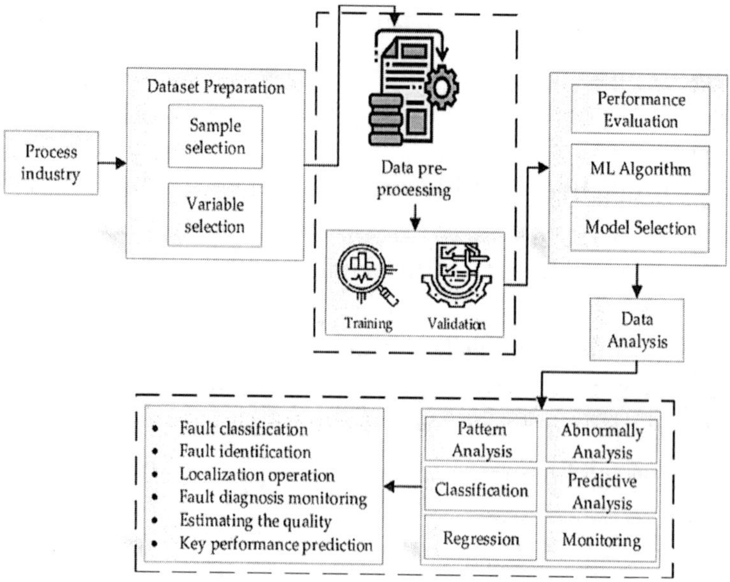

Fig 4.1 : ML-based fault detection process

Applications of Fault Detection in Various Industries

1. **Manufacturing** – Predictive maintenance in industrial automation prevents equipment failures.
2. **Telecommunications** – Network monitoring systems detect faults in communication networks.
3. **Automotive** – AI-driven fault detection ensures vehicle safety by monitoring sensors and electronic control units.
4. **Aerospace** – ML models analyze aircraft sensor data to predict

system failures.

Machine learning is revolutionizing fault detection by making systems smarter, reducing downtime, and improving efficiency across industries. As AI models become more sophisticated, fault detection systems will become even more reliable and proactive in preventing failures.

4.3 Machine Learning in Semiconductor Manufacturing

Semiconductor manufacturing is one of the most complex and precise industries, requiring high levels of accuracy, efficiency, and yield optimization. As semiconductor devices become smaller and more advanced, traditional process control methods face limitations in managing the vast amounts of data generated during fabrication. Machine learning (ML) is revolutionizing semiconductor manufacturing by enhancing defect detection, predictive maintenance, process optimization, and yield improvement through data-driven decision-making.

A. Role of Machine Learning in Semiconductor Manufacturing

Machine learning plays a crucial role in various stages of semiconductor manufacturing, including wafer fabrication, photolithography, chemical-mechanical polishing (CMP), etching, deposition, and testing. ML techniques analyze large datasets from sensors, equipment logs, and production systems to detect patterns, identify defects, and optimize process parameters in real time.

B. Key Applications of Machine Learning in Semiconductor Manufacturing

1. Defect Detection and Classification

Wafer defects and process variations can significantly impact chip performance. ML models, particularly deep learning-based image recognition techniques, are used to analyze semiconductor wafer inspection images and classify defects with high precision. Convolutional Neural Networks (CNNs) and Autoencoders are widely used for anomaly detection in wafer images, helping manufacturers reduce yield loss.

2. Predictive Maintenance

Machine learning enables predictive maintenance by analyzing equipment sensor data to detect early signs of failures. Traditional maintenance methods rely on fixed schedules or reactive approaches, leading to unnecessary downtime or unexpected failures. ML algorithms such as Support Vector Machines (SVM), Random Forests, and Recurrent Neural Networks (RNNs) predict potential equipment malfunctions, allowing manufacturers to perform maintenance before breakdowns occur, thereby reducing costs and improving efficiency.

3. Yield Optimization

Manufacturing yield is a key metric in semiconductor production. ML models analyze process parameters, environmental conditions, and material variations to identify factors that influence yield. By applying optimization algorithms like Bayesian Optimization and Gradient Boosting, semiconductor companies can fine-tune their processes to maximize output and minimize defective chips.

4. Process Control and Optimization

Semiconductor fabrication involves hundreds of process steps, and even minor deviations can lead to defects. ML models help optimize process parameters in real time by analyzing historical data and predicting the best settings for etching, deposition, and lithography processes. Reinforcement Learning (RL) techniques are used to adjust manufacturing conditions dynamically to achieve optimal production efficiency.

5. Supply Chain and Inventory Management

The semiconductor supply chain is highly complex, with dependencies on raw materials, production capacity, and logistics. ML-powered demand forecasting models help manufacturers optimize inventory management, reduce lead times, and minimize disruptions in production.

C. Future of Machine Learning in Semiconductor Industry

As semiconductor fabrication moves toward smaller nodes (e.g., 3nm and beyond), machine learning will become even more critical in managing manufacturing complexities. Future advancements will focus on AI-powered autonomous fabs, where real-time data-driven decisions continuously optimize production processes with minimal human intervention. Integration with edge AI, digital twins, and cloud computing will further enhance the efficiency and reliability of semiconductor manufacturing.

D. Machine learning is not just improving semiconductor production-it is revolutionizing the industry by making manufacturing smarter, more precise, and highly efficient.

4.4 Predictive Maintenance for Electronic Devices

Predictive maintenance (PdM) is a proactive approach that utilizes machine learning (ML) and artificial intelligence (AI) to anticipate equipment failures before they occur. Unlike traditional maintenance strategies—such as reactive maintenance (fixing issues after failure) or preventive maintenance (servicing at scheduled intervals)—predictive maintenance relies on real-time data analysis to determine the optimal time for maintenance.

In electronic devices, predictive maintenance is crucial for preventing unexpected failures, reducing downtime, and increasing the lifespan of components in industries such as manufacturing, telecommunications, healthcare, and consumer electronics. By continuously monitoring sensor data, ML models can detect anomalies, predict faults, and recommend necessary actions, leading to cost savings and improved reliability.

A. How Predictive Maintenance Works

Predictive maintenance in electronic devices typically follows these steps:

1. Data Collection

Sensors and monitoring tools collect real-time data on various parameters such as voltage, current, temperature, vibration, and usage patterns. This data is stored in cloud-based or edge-computing systems for further analysis.

2. Data Preprocessing

The collected raw data is cleaned, filtered, and transformed to remove noise and extract meaningful features. Techniques such as normalization and feature selection help improve the accuracy of ML models.

3. Anomaly Detection

Machine learning models analyze historical and real-time data to identify deviations from normal behavior. Unsupervised learning techniques, such as clustering and autoencoders, help detect unusual patterns that indicate potential failures.

4. Fault Prediction

Supervised learning algorithms, such as decision trees, support vector machines (SVM), and neural networks, predict when a failure is likely to occur. These models are trained on labeled datasets containing past failures and normal operating conditions.

5. Maintenance Scheduling

Once a potential failure is detected, predictive maintenance systems generate alerts and recommend maintenance actions. These actions can be automated or manually reviewed by engineers to prevent device failure before it happens.

B. Key Machine Learning Techniques Used in Predictive Maintenance

1. **Supervised Learning** – Algorithms like Random Forests and Gradient Boosting predict failures based on labeled historical data.
2. **Unsupervised Learning** – Clustering methods such as K-Means and Autoencoders detect anomalies without prior knowledge of failure patterns.
3. **Deep Learning** – Convolutional Neural Networks (CNNs) and

Recurrent Neural Networks (RNNs) analyze sensor data and time-series trends to improve predictive accuracy.

4. **Reinforcement Learning** – Adaptive models improve over time by learning the best maintenance strategies based on real-world feedback.

C. Benefits of Predictive Maintenance for Electronic Devices

1. **Minimized Downtime** – Early fault detection helps prevent unexpected device failures, improving system reliability.
2. **Cost Savings** – Reduces unnecessary maintenance and lowers repair costs by addressing issues before they escalate.
3. **Extended Device Lifespan** – Proactive maintenance ensures optimal device performance, preventing excessive wear and tear.
4. **Improved Safety** – Preventing failures in critical systems (e.g., medical devices, power grids) enhances safety and reduces risks.
5. **Data-Driven Decision Making** – ML-powered insights help manufacturers and maintenance teams optimize their processes efficiently.

D. Applications of Predictive Maintenance in Various Industries

1. **Consumer Electronics** – Smart home devices, laptops, and smartphones use ML-based maintenance to detect hardware degradation.
2. **Industrial Equipment** – Sensors in factory machines predict motor failures and overheating issues before they occur.
3. **Telecommunications** – Network infrastructure, such as routers and base stations, leverages ML for proactive maintenance.
4. **Healthcare Devices** – Medical equipment, including MRI scanners and ventilators, use predictive maintenance to ensure continuous operation.
5. **Automotive Electronics** – Predictive maintenance in electric vehicles (EVs) monitors battery health and component performance.

E. Future of Predictive Maintenance in Electronic Devices

As AI and IoT (Internet of Things) technology continue to advance, predictive maintenance will become even more sophisticated. Future developments will include:

- **Edge AI for Real-Time Monitoring** – Processing data locally on devices for instant fault detection without relying on cloud servers.
- **Digital Twins** – Virtual models of electronic devices that simulate real-world performance to predict failures more accurately.
- **Autonomous Maintenance Systems** – AI-driven self-repairing systems that automatically adjust settings or initiate self-healing mechanisms.

Machine learning-powered predictive maintenance is revolutionizing the way electronic devices are monitored and serviced, leading to increased efficiency, reduced costs, and enhanced reliability across industries.

4.5 ML for Signal Processing and Noise Reduction

Signal processing is essential in various applications, including telecommunications, biomedical engineering, speech recognition, radar systems, and image enhancement. One of the biggest challenges in signal processing is the presence of noise, which can distort useful information and degrade system performance. Traditional noise reduction methods, such as Fourier transforms, wavelet denoising, and Kalman filtering, have limitations in handling complex, non-stationary noise patterns.

Machine learning (ML) has revolutionized signal processing by providing adaptive and intelligent approaches for noise reduction, feature extraction, and signal enhancement. Unlike traditional techniques, ML algorithms can learn from data, adapt to changing noise environments, and improve performance across different applications.

A. Role of Machine Learning in Signal Processing

Machine learning plays a crucial role in improving signal processing efficiency by learning patterns and distinguishing between useful signals and noise. By training on large datasets, ML models can automatically filter out unwanted disturbances, enhance signal clarity, and improve decision-making in real time.

B.Key Applications of ML in Signal Processing

1. Noise Reduction and Denoising

ML techniques help remove background noise from signals while preserving important features. Deep learning models, such as Convolutional Neural Networks (CNNs) and Autoencoders, can analyze and reconstruct signals, eliminating interference more effectively than traditional filters.

2. Speech and Audio Enhancement

In voice communication systems, hearing aids, and smart assistants, ML-based speech enhancement improves voice clarity by suppressing background noise. Recurrent Neural Networks (RNNs) and Generative Adversarial Networks (GANs) are widely used for real-time speech processing.

3. Biomedical Signal Processing

Medical devices use ML to process physiological signals such as electrocardiograms (ECG), electroencephalograms (EEG), and electromyograms (EMG). These techniques help remove artifacts, detect abnormalities, and improve diagnostic accuracy.

4. Image and Video Noise Reduction

ML-based signal processing enhances image and video quality by reducing noise and improving resolution. Deep learning architectures like U-Net and GANs are used for denoising and super-resolution applications in photography, medical imaging, and security systems.

5. Radar and Wireless Communication

In radar systems and wireless networks, ML algorithms optimize signal transmission by filtering out environmental noise, detecting weak signals, and improving communication reliability. Reinforcement Learning (RL) is often applied to dynamically adjust signal parameters.

C. Machine Learning Techniques for Noise Reduction

1.Supervised Learning – Uses labeled datasets of noisy and clean signals to train models for noise reduction. Common algorithms include Support Vector Machines (SVMs) and Deep Neural Networks (DNNs).

2. Unsupervised Learning – Identifies and separates noise without labeled data using techniques like Principal Component Analysis (PCA) and clustering methods.

3. Deep Learning – CNNs, Autoencoders, and Long Short-Term Memory (LSTM) networks extract features and filter noise efficiently in complex signals.

4. Reinforcement Learning – Adaptive models improve noise reduction strategies over time by optimizing system performance through feedback mechanisms.

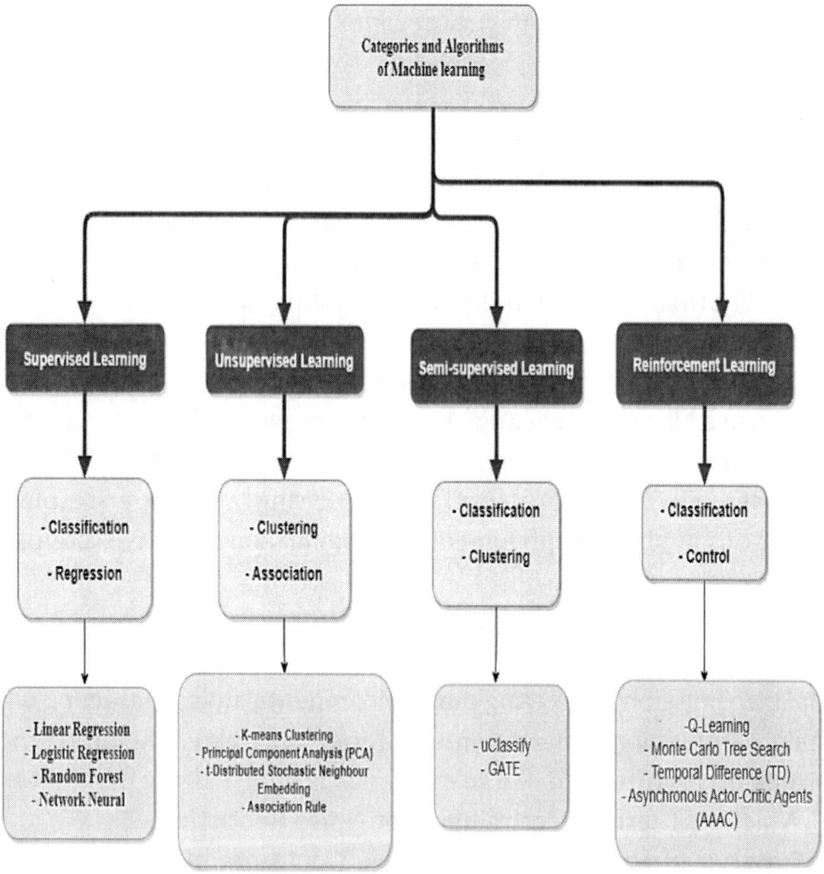

Fig 4.2: ML-based signal processing workflow

D. Benefits of Machine Learning in Signal Processing

1. Improved Noise Suppression – ML algorithms adapt to different noise environments for better denoising.

2. Real-Time Processing – AI-powered noise reduction enhances real-time signal quality in various applications.

3. Higher Accuracy – ML-driven processing improves speech recognition, medical diagnostics, and wireless communication.

4. Scalability – AI-based noise reduction is applicable to multiple signal types, from audio and images to biomedical and radar signals.

E. Future of Machine Learning in Signal Processing

The integration of machine learning with edge computing, IoT, and 5G networks will further enhance real-time signal processing applications. Future advancements may include:

- **Self-Adaptive Noise Reduction Systems** – AI models dynamically adjusting to changing noise environments.
- **Quantum Machine Learning for Signal Processing** – Next-generation algorithms leveraging quantum computing for ultra-fast signal analysis.
- **AI-Powered Brain-Computer Interfaces (BCI)** – Advanced signal processing techniques enabling direct communication between human brains and machines.

Machine learning is transforming signal processing by making systems more adaptive, accurate, and efficient, paving the way for technological advancements across multiple industries.

4.6 Edge AI and Embedded Machine Learning

Edge AI and Embedded Machine Learning are revolutionizing how intelligent systems process data in real time. Traditional AI models rely on cloud computing for data processing, which can introduce latency, security concerns, and high bandwidth costs. Edge AI moves AI computations closer to the source of data—on edge devices such as IoT sensors, mobile devices, and embedded systems. Embedded Machine Learning (Embedded ML) further enhances this by enabling machine learning models to run efficiently on low-power, resource-constrained hardware.

Together, Edge AI and Embedded ML enable real-time decision-making, reduced dependence on cloud servers, improved data privacy, and enhanced efficiency across industries such as healthcare, automotive, manufacturing, and smart cities.

A. What is Edge AI?

Edge AI refers to deploying artificial intelligence algorithms directly on edge devices rather than sending data to a central cloud server for processing. This allows AI-powered applications to function in real time, reducing latency and improving responsiveness. Edge AI is commonly implemented in:

- **Smartphones and Wearables** – AI-powered voice assistants, facial recognition, and health tracking.
- **Autonomous Vehicles** – Real-time object detection and navigation.
- **Industrial IoT (IIoT) Devices** – Predictive maintenance and automation in manufacturing.
- **Smart Cameras and Security Systems** – AI-based surveillance with on-device anomaly detection.

Advantages of Edge AI

- **Low Latency** – AI computations occur locally, enabling real-time decision-making.
- **Enhanced Privacy** – Sensitive data remains on the device, reducing security risks.
- **Reduced Cloud Dependency** – Lower bandwidth usage and operational costs.
- **Improved Reliability** – Devices can operate even in limited or no-network environments.

B. What is Embedded Machine Learning?

Embedded ML refers to running machine learning models on embedded systems—small, specialized hardware with limited processing power, such as microcontrollers (MCUs), single-board computers (e.g., Raspberry Pi), and AI accelerators (e.g., Google Edge TPU). These systems can process data efficiently with minimal energy consumption, making them ideal for battery-powered and real-time applications.

Applications of Embedded ML

1. **Healthcare Devices** – AI-powered ECG monitoring, smart prosthetics, and glucose monitoring.
2. **Smart Home Devices** – Voice recognition, energy-efficient smart appliances, and gesture-based controls.
3. **Agriculture and Environmental Monitoring** – AI-powered sensors detecting soil conditions, weather patterns, and crop health.
4. **Industrial Automation** – Embedded AI in robots and sensors for predictive maintenance and quality control.
5. **Autonomous Drones** – Real-time AI processing for object tracking, navigation, and surveillance.

C. Challenges of Embedded ML

- **Hardware Constraints** – Limited processing power, memory, and battery life.
- **Model Optimization** – Requires techniques such as quantization and pruning to reduce model size.
- **Real-Time Performance** – Efficient execution of ML models on low-power hardware.

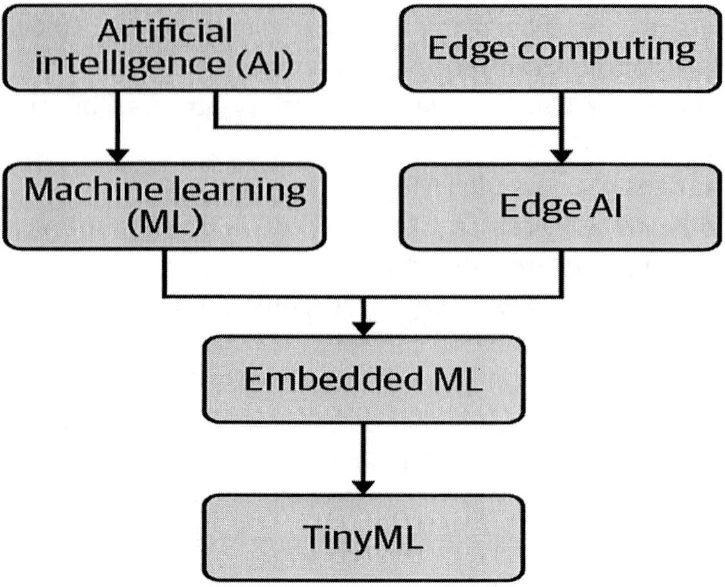

Fig 4.3: diagram illustrating the workflow of Edge AI and Embedded ML

D. Technologies Enabling Edge AI and Embedded ML

Several hardware and software innovations are driving the adoption of Edge AI and Embedded ML:

1. AI Accelerators

- **Google Edge TPU** – Specialized AI chip for real-time inference.
- **NVIDIA Jetson** – AI platform for robotics and edge computing.
- **Intel Movidius** – AI vision processing unit for low-power inference.

2. Frameworks for Embedded ML

- **TensorFlow Lite** – Optimized ML models for mobile and embedded devices.
- **TinyML** – Ultra-low-power ML models running on microcontrollers.
- **Edge Impulse** – End-to-end platform for developing embedded ML applications.

E. Benefits of Edge AI and Embedded ML

1. **Real-Time Decision Making** – AI models run directly on the

device, ensuring instant responses.

2. **Energy Efficiency** – Optimized models consume minimal power, ideal for battery-operated devices.
3. **Scalability** – AI-powered embedded systems can be deployed across industries at scale.
4. **Cost Savings** – Reduced cloud processing and bandwidth costs improve operational efficiency.

F. Future of Edge AI and Embedded Machine Learning

As AI hardware advances, Edge AI and Embedded ML will continue to evolve with:

- **Ultra-Low Power AI Chips** – AI accelerators designed for extreme efficiency.
- **Federated Learning on Edge** – Privacy-preserving AI training across multiple devices without centralizing data.
- **5G and AI Synergy** – Faster data transmission enabling smarter edge applications.
- **Self-Learning AI Systems** – Devices that continuously improve without cloud dependency.

Edge AI and Embedded ML are transforming industries by enabling AI-powered intelligence at the device level, paving the way for smarter, more efficient, and highly autonomous systems.

Chapter 5

Machine Learning in Electrical Engineering

5.1 Smart Grid and Energy Management Systems

The global demand for electricity is rising, necessitating more efficient, reliable, and sustainable power distribution. Traditional power grids are increasingly struggling with issues like energy losses, inefficiencies, and outages. The Smart Grid is an advanced energy infrastructure that integrates digital technology, automation, and real-time data analytics to improve energy generation, distribution, and consumption.

At the core of the Smart Grid lies Energy Management Systems (EMS), which optimize energy flow, reduce waste, and enhance grid reliability. By leveraging technologies such as Artificial Intelligence (AI), Internet of Things (IoT), and Machine Learning (ML), these systems enable intelligent decision-making, demand-side management, and integration of renewable energy sources.

A. What is a Smart Grid?

A Smart Grid is an intelligent electricity network that incorporates modern computing and communication technologies to enhance the monitoring, control, and optimization of power generation and distribution. Unlike traditional grids, Smart Grids are bidirectional, meaning they allow energy flow both from producers to consumers and from consumers back to the grid (e.g., solar power contributions).

B. Key Features of Smart Grids

- **Automated Grid Monitoring** – Sensors and smart meters provide real-time data on energy consumption and faults.
- **Demand Response Management** – Adjusts electricity demand based on grid conditions, reducing peak loads.
- **Renewable Energy Integration** – Efficiently incorporates

solar, wind, and other renewable sources into the grid.

- **Self-Healing Capabilities** – Detects faults and reroutes electricity to prevent outages.
- **Decentralized Energy Management** – Consumers (prosumers) can generate and share excess energy.

C. Applications of Smart Grids

1. **Smart Metering** – Real-time tracking of energy usage for better efficiency.
2. **Distributed Energy Resources (DERs)** – Integration of local solar panels and wind turbines.
3. **Electric Vehicle (EV) Charging Management** – Smart charging stations optimize energy use.
4. **Grid Automation** – AI-based control systems detect and correct power failures.

D. What is an Energy Management System (EMS)?

An **Energy Management System (EMS)** is a technology-driven system that monitors, controls, and optimizes energy consumption across homes, industries, and power grids. It provides **real-time analytics, predictive maintenance, and automation** to improve energy efficiency and sustainability.

Types of Energy Management Systems

1. **Home Energy Management Systems (HEMS)** – Smart home devices optimize electricity usage.
2. **Building Energy Management Systems (BEMS)** – AI-powered solutions improve efficiency in commercial buildings.
3. **Industrial Energy Management Systems (IEMS)** – Large-scale energy monitoring in factories and industrial plants.
4. **Utility-Level EMS** – Grid operators use AI-based tools for demand response and outage prevention.

E. Technologies Enabling Smart Grids and EMS

1. **Internet of Things (IoT)** – Smart sensors and meters collect real-time data.

2. **Artificial Intelligence (AI) & Machine Learning (ML)** – Predictive analytics optimize energy distribution.
3. **Blockchain Technology** – Secures energy transactions in decentralized power networks.
4. **Cloud Computing** – Stores and processes massive amounts of energy data for insights.
5. **Edge Computing** – Enables real-time energy optimization at local levels.

F. Benefits of Smart Grids and EMS

- **Improved Energy Efficiency** – Reduces energy waste and optimizes electricity usage.
- **Lower Costs for Consumers** – Demand response programs offer lower rates during off-peak hours.
- **Better Grid Reliability** – AI-powered monitoring detects and prevents failures.
- **Seamless Integration of Renewables** – Smart grids manage fluctuating solar and wind energy efficiently.
- **Reduced Carbon Footprint** – Optimized energy usage leads to lower greenhouse gas emissions.

G. Future of Smart Grids and EMS

As the world transitions to a **sustainable energy future**, Smart Grids and EMS will play a key role in:

- **AI-Powered Autonomous Grids** – Self-learning systems that optimize energy flow in real time.
- **Wireless Energy Transfer** – Future grids may incorporate contactless power distribution.
- **Smart City Integration** – AI-based energy optimization for urban infrastructures.
- **Energy Trading Marketplaces** – Decentralized blockchain-powered peer-to-peer energy trading.

Smart Grids and EMS are transforming how we generate, distribute, and consume energy, paving the way for a smarter, more sustainable, and resilient power infrastructure.

5.2 Machine Learning for Power Load Forecasting

Power load forecasting is a critical aspect of energy management, enabling utilities, industries, and consumers to predict electricity demand accurately. Traditional forecasting methods rely on statistical techniques, which often fail to capture complex, non-linear patterns in energy consumption. **Machine Learning (ML) is revolutionizing power load forecasting** by leveraging large datasets, real-time analytics, and advanced algorithms to improve accuracy, efficiency, and reliability.

By applying ML, energy providers can **optimize power generation, reduce costs, enhance grid stability, and integrate renewable energy sources** more effectively. This technology helps predict electricity demand at different time scales—short-term (hours to days), medium-term (weeks to months), and long-term (years).

A. What is Power Load Forecasting?

Power load forecasting refers to predicting future electricity demand based on historical data, weather patterns, economic indicators, and consumer behavior. Accurate forecasting ensures optimal energy distribution, prevents blackouts, and reduces operational costs for power companies.

Types of Load Forecasting

1. **Short-Term Load Forecasting (STLF)** – Predicts demand within hours or days, aiding real-time grid management.
2. **Medium-Term Load Forecasting (MTLF)** – Forecasts demand over weeks or months for planning maintenance and fuel supply.
3. **Long-Term Load Forecasting (LTLF)** – Estimates electricity demand for years ahead to support infrastructure development.

Challenges in Traditional Load Forecasting

- **Variability in Energy Demand** – Consumption fluctuates due to human activities, seasonal changes, and economic shifts.
- **Incorporation of Renewable Energy** – Solar and wind energy introduce unpredictable fluctuations.
- **Lack of Real-Time Adaptability** – Traditional models struggle to process real-time data effectively.

B. Machine Learning in Power Load Forecasting

Machine Learning enhances forecasting by **identifying hidden patterns, handling large datasets, and continuously improving predictions** through learning algorithms.

Key ML Techniques for Load Forecasting

1. **Regression Models** – Linear and polynomial regression analyze historical energy consumption trends.
2. **Time Series Models** – ARIMA, LSTMs, and RNNs capture sequential dependencies in energy demand.
3. **Neural Networks (NNs)** – Deep learning models like CNNs and LSTMs improve accuracy for complex data patterns.
4. **Support Vector Machines (SVMs)** – Classifies and predicts power demand with high precision.
5. **Ensemble Learning** – Combines multiple models (e.g., Random Forest, Gradient Boosting) for robust predictions.

C. Benefits of ML-Based Load Forecasting

- **Higher Accuracy** – ML algorithms outperform traditional statistical models.
- **Real-Time Predictions** – AI models continuously learn and adapt to real-time changes.
- **Improved Energy Efficiency** – Accurate forecasts prevent overproduction and reduce energy waste.
- **Cost Reduction** – Optimized power generation minimizes fuel consumption and operational costs.
- **Seamless Renewable Energy Integration** – AI models predict fluctuations in solar and wind energy availability.

D. Future of Machine Learning in Load Forecasting

As energy systems become smarter, ML-driven load forecasting will play a crucial role in:

- **AI-Powered Smart Grids** – Self-learning grids optimizing energy distribution dynamically.
- **Edge AI for Localized Predictions** – AI models running on local devices for decentralized forecasting.

- **Quantum Computing in Load Prediction** – Advanced computing for ultra-fast energy forecasting.
- **AI-Driven Demand Response Systems** – Automated load balancing for enhanced energy efficiency.

Machine Learning is **redefining power load forecasting**, making electricity distribution more reliable, efficient, and adaptable to future energy demands.

5.3 Fault Diagnosis in Electrical Systems

Electrical systems are the backbone of industries, transportation, and daily life. Ensuring their reliability is critical to preventing power failures, equipment damage, and safety hazards. **Fault diagnosis** is the process of detecting, classifying, and predicting electrical faults before they cause serious failures. Traditional fault diagnosis methods rely on manual inspection, rule-based systems, and signal analysis, which can be slow and inefficient.

A. Machine Learning (ML) is revolutionizing fault diagnosis by leveraging data-driven models that automatically detect patterns, predict failures, and optimize maintenance strategies. With real-time monitoring and predictive analytics, ML-based fault diagnosis enhances system reliability, reduces downtime, and minimizes maintenance costs.

C.What is Fault Diagnosis in Electrical Systems?

Fault diagnosis involves identifying electrical faults—such as **short circuits, insulation failures, and power surges**—in power grids, industrial machines, and household appliances. The goal is to detect faults early, determine their causes, and recommend corrective actions.

Types of Electrical Faults

1. **Short Circuit Faults** – Sudden increase in current due to unintended connections.
2. **Open Circuit Faults** – Broken wires or disconnected components causing a loss of power.
3. **Ground Faults** – Leakage of electrical current to the ground,

leading to potential hazards.

4. **Transient Faults** – Temporary disturbances like lightning strikes or switching surges.
5. **Overload Faults** – Excessive power demand leading to overheating and equipment failure.

D. Challenges in Traditional Fault Diagnosis

- **Manual Inspection** – Time-consuming and prone to human error.
- **Complexity of Electrical Systems** – Multiple fault types make diagnosis difficult.
- **Lack of Predictive Insights** – Traditional methods focus on reactive rather than proactive maintenance.

E. Machine Learning in Fault Diagnosis

Machine Learning (ML) improves fault diagnosis by analyzing historical data, real-time sensor readings, and electrical waveforms to detect anomalies and predict failures.

E. Key ML Techniques for Fault Diagnosis

1. **Supervised Learning** – Uses labeled datasets to train models to classify different fault types. (e.g., Support Vector Machines, Random Forests)
2. **Unsupervised Learning** – Detects anomalies and hidden patterns without predefined labels. (e.g., Clustering, Autoencoders)
3. **Deep Learning** – Neural networks extract features from electrical signals to enhance accuracy. (e.g., Convolutional Neural Networks, Recurrent Neural Networks)
4. **Reinforcement Learning** – AI-based decision-making for real-time fault detection and corrective actions.

F. Benefits of ML-Based Fault Diagnosis

1. **Early Fault Detection** – Identifies faults before they escalate into major failures.
2. **Real-Time Monitoring** – AI models analyze sensor data continuously for instant detection.

3. **Reduced Downtime** – Predictive maintenance prevents unexpected equipment shutdowns.
4. **Cost Savings** – Minimizes repair costs by addressing issues proactively.
5. **Enhanced Safety** – Detecting faults early reduces risks of electrical hazards.

G. Future of Machine Learning in Fault Diagnosis

With advancements in AI and IoT, ML-driven fault diagnosis will continue to evolve with:

- **AI-Powered Self-Healing Systems** – Automated corrective actions based on real-time diagnostics.
- **Edge Computing for Fault Detection** – AI models running locally on embedded devices for faster response.
- **Digital Twins** – Simulated models of electrical systems for predictive fault analysis.
- **Federated Learning for Secure Data Processing** – Distributed AI models ensuring privacy and efficiency.
- Machine Learning is transforming fault diagnosis in electrical systems, making them **smarter, more reliable, and safer** for the future.

5.4. Predictive Maintenance of Electrical Equipment

Electrical equipment plays a crucial role in industries, power grids, and daily life. However, electrical failures can lead to costly downtime, production losses, and safety hazards. **Predictive maintenance (PdM)** is a proactive approach that leverages data analytics and machine learning (ML) to **anticipate equipment failures before they occur**. Unlike traditional reactive or preventive maintenance, predictive maintenance minimizes unnecessary servicing, reduces operational costs, and enhances equipment lifespan.

With advancements in **Artificial Intelligence (AI), the Internet of Things (IoT), and cloud computing**, predictive maintenance has become more efficient, enabling real-time monitoring and intelligent decision-making. This approach allows companies to move away from scheduled maintenance cycles and instead **predict failures based on**

real-world data.

A. What is Predictive Maintenance?

Predictive maintenance involves using **sensor data, machine learning algorithms, and predictive analytics** to detect early signs of faults in electrical equipment. By analyzing historical data, PdM models can **identify patterns, diagnose anomalies, and estimate the remaining useful life (RUL) of equipment**.

B. Types of Maintenance Approaches

1. **Reactive Maintenance** – Fixing equipment after failure (high downtime and costs).
2. **Preventive Maintenance** – Scheduled servicing based on time intervals (may lead to unnecessary maintenance).
3. **Predictive Maintenance** – AI-driven fault prediction based on actual equipment conditions (minimizes downtime and costs).

C. Challenges in Traditional Maintenance

- **Unnecessary Service Costs** – Preventive maintenance may replace components too early.
- **Unexpected Failures** – Reactive maintenance leads to costly breakdowns and downtime.
- **Lack of Real-Time Monitoring** – Manual inspections fail to detect early-stage failures.

D. Machine Learning in Predictive Maintenance

Machine Learning enhances predictive maintenance by analyzing real-time data from **sensors, IoT devices, and historical logs** to detect faults before they cause major failures.

E. Key ML Techniques for Predictive Maintenance

1. **Supervised Learning** – Classification models (Random Forest, SVM) detect known failure patterns.
2. **Unsupervised Learning** – Anomaly detection models (Autoencoders, K-Means Clustering) identify hidden faults.

3. **Time-Series Analysis** – Recurrent Neural Networks (RNNs) and Long Short-Term Memory (LSTM) models predict equipment degradation over time.
4. **Deep Learning** – Convolutional Neural Networks (CNNs) analyze waveform and vibration data for fault detection.

F. Benefits of ML-Based Predictive Maintenance

- **Early Fault Detection** – Identifies issues before they cause major failures.
- **Reduced Downtime** – Prevents unplanned outages and improves productivity.
- **Optimized Maintenance Costs** – Reduces unnecessary servicing and component replacements.
- **Enhanced Equipment Lifespan** – Ensures machines operate efficiently for longer.
- **Real-Time Monitoring** – AI-powered analytics provide instant insights on equipment health.

E. Future of Predictive Maintenance

With ongoing advancements in **AI, IoT, and cloud computing**, the future of predictive maintenance includes:

- **AI-Driven Automated Repairs** – Intelligent robots and self-healing systems.
- **Edge AI for Localized Processing** – Real-time fault detection at the device level.
- **Digital Twins** – Virtual simulations of electrical equipment for predictive analytics.
- **Federated Learning** – Secure AI models that process decentralized maintenance data.

Machine Learning is **transforming predictive maintenance**, ensuring electrical equipment operates reliably, efficiently, and with minimal disruptions.

5.5. Machine Learning in Renewable Energy Systems

Renewable energy systems—such as **solar, wind, hydro, and**

biomass power—play a vital role in reducing dependence on fossil fuels and mitigating climate change. However, these energy sources are highly variable and dependent on **weather conditions, geographic factors, and grid demands**. This unpredictability creates challenges in energy generation, distribution, and storage.

Machine Learning (ML) is transforming renewable energy systems by enabling accurate forecasting, optimizing power generation, and improving grid integration. Through **real-time data analysis, pattern recognition, and predictive modeling**, ML enhances efficiency, reduces costs, and ensures a stable power supply.

A. Challenges in Renewable Energy Systems

- **Intermittency** – Solar and wind power depend on environmental factors, making them unpredictable.
- **Energy Storage Optimization** – Efficient battery usage is crucial for grid stability.
- **Grid Integration** – Balancing energy supply and demand requires smart forecasting.
- **Equipment Maintenance** – Predictive maintenance ensures smooth operation of solar panels, wind turbines, and hydroelectric plants.

B. Applications of Machine Learning in Renewable Energy

1. Renewable Energy Forecasting

Solar Power Prediction – ML models analyze satellite images, weather data, and past generation patterns to predict solar power output.

Wind Power Forecasting – AI-based models predict wind speed and direction to optimize turbine efficiency.

2. Smart Grid Management

Demand-Supply Balancing – ML algorithms adjust energy distribution in real-time based on demand fluctuations.

Energy Storage Optimization – AI-powered battery management maximizes storage efficiency and lifespan.

3. Predictive Maintenance for Renewable Energy Equipment

Wind Turbine Fault Detection – ML models analyze vibration, temperature, and pressure data to detect early faults.

Solar Panel Health Monitoring – AI-based image recognition detects dirt accumulation, shading, and cell degradation.

C. Benefits of ML in Renewable Energy Systems

- **Improved Forecast Accuracy** – AI-driven predictions enhance energy planning and stability.
- **Higher Energy Efficiency** – Optimized resource allocation reduces waste and costs.
- **Real-Time Grid Optimization** – AI balances supply, demand, and storage automatically.
- **Lower Maintenance Costs** – Predictive analytics prevent unexpected equipment failures..
- **Better Renewable Integration** – AI enables seamless adoption of solar and wind power into the grid.

D. Future of Machine Learning in Renewable Energy

As AI continues to evolve, **renewable energy systems will become smarter and more efficient**. Future advancements include:

- **AI-Powered Smart Grids** – Self-learning energy systems optimizing power flow dynamically.
- **Edge Computing for Energy Management** – Real-time AI processing at power plants and grid nodes.
- **Blockchain for Decentralized Energy Trading** – AI-powered secure peer-to-peer energy transactions.
- **AI-Optimized Carbon Capture** – Intelligent systems for monitoring and reducing CO_2 emissions.

Machine Learning is **revolutionizing renewable energy**, making it more reliable, cost-effective, and sustainable for the future.

5.6. Robotics and Automation in Electrical Engineering

Electrical engineering is rapidly evolving with the integration of **robotics and automation**, leading to smarter, more efficient systems in

power generation, manufacturing, maintenance, and energy distribution. Robotics and automation play a crucial role in enhancing productivity, improving safety, and reducing operational costs in industries such as **power plants, renewable energy systems, smart grids, and electrical maintenance**.

With advancements in **Artificial Intelligence (AI), the Internet of Things (IoT), and Machine Learning (ML)**, automated electrical systems can now operate with minimal human intervention, making processes **faster, more precise, and more reliable**.

A. Applications of Robotics and Automation in Electrical Engineering

1. Automated Power Grid Management

Smart Grid Optimization – AI-driven automation manages electricity distribution efficiently.

Fault Detection and Repair – Drones and robotic systems inspect and repair power lines.

2. Industrial Automation

Automated Circuit Assembly – Robotics improve accuracy in PCB and microchip manufacturing.

Robotic Process Automation (RPA) – AI-driven software automates energy management task.

3. Predictive Maintenance

Transformer and Substation Monitoring – Automated robots detect faults before failures occur.

AI-Based Electrical Equipment Diagnostics – Machine learning analyzes equipment health.

4. Renewable Energy Automation

Solar Panel Cleaning Robots – AI-powered robots ensure optimal energy generation.

Wind Turbine Maintenance Drones – Drones inspect and repair turbine blades autonomously.

B.Benefits of Robotics and Automation in Electrical Engineering

- **Increased Efficiency** – Robotics enhance productivity by automating repetitive electrical tasks.
- **Improved Safety** – AI-driven robots reduce human exposure to electrical hazards.
- **Cost Savings** – Automation reduces labor costs and minimizes energy wastage.
- **Enhanced Accuracy** – AI-powered robots eliminate human errors in circuit design and maintenance.
- **Real-Time Monitoring** – IoT-enabled automation provides instant updates on system performance.

C.Future of Robotics and Automation in Electrical Engineering

As technology advances, robotics and automation will continue to shape electrical engineering in areas such as:

- **AI-Driven Self-Healing Grids** – Automated systems detecting and fixing power issues in real-time.
- **Edge Computing for Faster Automation** – Decentralized AI processing for instant system responses.
- **Autonomous Electrical Maintenance Robots** – AI-powered bots for transformer and substation inspections.
- **Quantum Computing for Electrical System Optimization** – High-speed problem-solving in circuit design and power distribution.

Robotics and automation are **revolutionizing electrical engineering**, making electrical systems smarter, safer, and more efficient for the future.

5.7. Internet of Things (IoT) and Machine Learning Applications

The **Internet of Things (IoT)** and **Machine Learning (ML)** are transforming industries by enabling real-time data collection, intelligent decision-making, and automation. IoT refers to a network of connected devices—such as **sensors, smart appliances, and industrial machines**—that communicate with each other and with cloud-based systems. Meanwhile, **Machine Learning** allows these devices to analyze data, detect patterns, and make predictions without human intervention.

When combined, **IoT and ML** create smarter systems that optimize energy consumption, enhance security, improve healthcare, and automate industrial processes. From **smart homes** to **predictive maintenance in factories**, these technologies are reshaping the future.

A. Applications of IoT and Machine Learning

1. Smart Homes and Smart Cities

Home Automation – AI-powered IoT devices control lighting, temperature, and security.

Traffic Optimization – Smart sensors analyze traffic flow and adjust signals dynamically.

2. Healthcare and Wearable Technology

Remote Patient Monitoring – AI-driven wearables track heart rate, oxygen levels, and activity.

Disease Prediction – Machine Learning models detect early signs of illnesses using IoT sensor data.

3. Industrial IoT (IIoT) and Smart Manufacturing

Predictive Maintenance – IoT sensors monitor machine health, and AI predicts failures.

Supply Chain Optimization – ML algorithms analyze demand patterns to streamline logistics.

4. Energy Management and Smart Grids

Smart Meters – IoT devices track electricity usage, and ML optimizes energy distribution.

Renewable Energy Forecasting – AI predicts solar and wind power generation for grid stability.

5. Security and Surveillance

AI-Powered Surveillance – IoT cameras detect anomalies and alert authorities in real-time.

Cybersecurity – ML algorithms identify suspicious network activities and prevent cyber threats.

B. Benefits of IoT and Machine Learning

- **Real-Time Decision Making** – AI-powered IoT devices analyze data instantly.
- **Increased Efficiency** – Automated systems reduce human workload and optimize operations.
- **Cost Savings** – Predictive maintenance and energy optimization reduce expenses.
- **Enhanced Security** – AI-driven cybersecurity protects IoT networks from cyber threats.
- **Scalability** – Cloud-based ML models handle massive amounts of IoT data efficiently.

C. Future of IoT and Machine Learning

As technology advances, IoT and ML will drive innovations in areas such as:

- **5G-Enabled IoT Networks** – Faster and more reliable IoT data transmission.
- **AI-Powered Edge Computing** – ML models running on local IoT devices for instant decision-making.
- **Blockchain for Secure IoT Transactions** – Decentralized data security for smart devices.
- **Autonomous Systems** – Self-learning IoT devices adapting to changing environments.

IoT and Machine Learning are **revolutionizing industries**, making the world **smarter, safer, and more connected** than ever before.

Chapter 6

Machine Learning Emerging Trends and Future Directions

Emerging Trends and Future Directions

In recent years, machine learning has developed and spread rapidly, revolutionizing various industries and fields. This chapter examines emerging trends and future directions in machine learning, highlighting cutting-edge techniques, algorithms, and applications that have tremendous potential to shape the field in the coming years. In addition, relevant figures and tables are presented in this chapter to improve the understanding and visualization of these trends.

A. Reinforcement learning and deep reinforcement learning:

Reinforcement learning (RL) has received considerable attention for its ability to train intelligent agents to interact with dynamic environments and optimize their actions based on rewards and punishments. RL has found applications in robotics, gaming and autonomous systems. Deep Reinforcement Learning (DRL), which combines deep learning with RL, has achieved significant success in solving complex problems, surpassing human-level performance in games such as Go and Dota 2.

Figure 1 illustrates the RL process, while Table 1 compares RL and Dota 2 .DRL: Ugh. Algorithms

Table 6.1: Comparison of RL and DRL Algorithms

Algorithm	Reinforcement Learning	Deep Reinforcement Learning
Q-Learning	Yes	Yes
Deep Q-Networks	No	Yes
Policy Gradient	Yes	Yes

Proximal Policy Optimization (PPO)	No	Yes
Actor-Critic	Yes	Yes

B. Transfer Learning and Meta-Learning:

Applying knowledge acquired from one task or topic to another has never been easier thanks to transfer learning, which has become a potent tool. Transfer learning lessens the requirement for significant labelled data in new domains by transferring learnt representations or models. This makes learning new tasks quicker and more effective. On the other hand, meta-learning focuses on learning how to learn. By utilising previous knowledge from a group of related jobs, it seeks to create models that can swiftly adapt to new tasks. Transfer learning is illustrated in Figure 2 while prominent meta-learning techniques are shown in Table 2.

Table 6.2: Popular Meta-Learning Algorithms

Algorithm	Description
Model-Agnostic Meta-Learning (MAML)	Learns an initialization that can be fine-tuned rapidly
Reptile	Performs gradient descent iterations on multiple tasks
Learning to Learn	Trains a recurrent neural network for task adaptation
Gradient-Based Meta-Learning (GBML)	Optimizes model parameters for fast adaptation

C. Federated learning and privacy-preserving machine learning: Federated learning is a distributed learning strategy that enables training models across several decentralized devices while protecting data privacy. With federated learning, devices can learn together while keeping their local data local rather than sending raw data to a central server. This method is especially useful in fields like healthcare and finance where data cannot be freely shared owing to privacy issues. The federated learning process is depicted in Figure 3, and various privacy-preserving machine learning algorithms are contrasted in Table 3.

Table 6.3: Comparison of Privacy-Preserving Machine Learning Techniques

Technique	Description
Homomorphic Encryption	Performs computations on encrypted data without decryption
Differential Privacy	Adds noise to the data to protect individual privacy
Secure Multi-Party Computation (SMPC)	Allows collaborative computation on encrypted data

D. Machine learning that can be explained and interpreted: As machine learning systems are increasingly used in crucial fields, it is essential that they can be explained and interpreted. Explainable machine learning (XAI) tries to give consumers understanding of the underlying decision-making process by offering insights into how models produce predictions. The goal of interpretable machine learning (IML) is to create models that are visible and comprehensible by nature. An illustration of a XAI technique is shown in Figure 4, and interpretable machine learning algorithms are highlighted in Table 4.

Table 6.4: Interpretable Machine Learning Algorithms

Algorithm	Description
Decision Trees	Creates hierarchical decision rules
Rule-based Models	Represents decisions as a set of IF-THEN rules
LIME (Local Interpretable Model-Agnostic Explanations)	Generates interpretable explanations for complex models

SHAP (Shapley Additive exPlanations)	Quantifies the impact of each feature on a model's output

Machine learning is a constantly developing field with a wide range of new trends and potential future directions. Reinforcement learning, transfer learning, federated learning, and explainable/interpretable machine learning are some of the important topics that have been covered in this chapter. Researchers and practitioners can open up new opportunities and promote the development of intelligent systems across diverse fields by embracing these developments and tackling the problems they provide. It is essential to stay up to date on the most recent developments in the area as it develops and to make adjustments to the machine learning environment as it changes.

6.1 Advances in Deep Learning and Neural Architectures

Deep learning has emerged as a powerful paradigm in the field of artificial intelligence, enabling significant breakthroughs in various fields, including computer vision, natural language processing, and speech recognition. This chapter reviews recent advances in deep learning and neural architectures and highlights key techniques and architectures that have contributed to state-of-the-art performance in these areas. In addition, related figures and tables are provided to improve understanding and visualization of these advances.

A. Convolutional Neural Networks (CNN):

Convolutional Neural Networks (CNNs) have revolutionized computer vision tasks, enabling excellent results in image classification, object detection and semantic segmentation. Recent advances in CNNs have focused on improving their efficiency, interpretability, and generalizability. Figure 1 illustrates a typical CNN architecture, while Table 1 outlines prominent CNN architectures and their applications.

Table 6.5: Notable CNN Architectures and Applications

Architecture	Applications
AlexNet	Image classification
VGG	Image classification, object detection
ResNet	Image classification, object detection
Inception	Image classification, object detection
DenseNet	Image classification, semantic segmentation

B. Recurrent Neural Networks (RNN) and Long Short Memory (LSTM):

Recurrent Neural Networks (RNNs) have helped model sequential data such as natural language processing and speech recognition. However, traditional RNNs suffer from the vanishing gradient problem, which limits their ability to capture long-term dependencies. Long Short-Term Memory (LSTM) networks, an extension of RNNs, solve this problem by involving memory cells. Figure 2 shows the network architecture of LSTM, while Table 2 shows various implementations of RNN and LSTM.

Table 6.6: Applications of RNNs and LSTMs

Application	Model
Natural Language Processing	LSTM
Speech Recognition	LSTM, Gated Recurrent Unit (GRU)
Music Generation	LSTM
Sentiment Analysis	RNN

C. Generative Adverbial Networks (GAN):

Generative adversarial networks (GANs) revolutionized the field of generative modeling by introducing a game theoretic framework. GANs consist of a generator network that learns to generate synthetic samples and a discriminator network that learns to distinguish between real and fake samples. The interaction between these networks leads to the generation of high-quality samples. Figure 3 shows the GAN architecture, while Table 3 outlines the major variants of GANs and their applications.

Table 6.7: Notable GAN Variants and Applications

GAN Variant	Applications
DCGAN (Deep Convolutional GAN)	Image generation
CycleGAN	Image-to-image translation, style transfer
StackGAN	Text-to-image synthesis
StarGAN	Facial attribute manipulation

D. Transformative networks:

Transformer networks have revolutionized the field of natural language processing, especially in machine translation and language understanding tasks. Unlike traditional iteration architectures, transformers use self-monitoring mechanisms to efficiently capture global dependencies. Transformers have achieved top-level performance in various language-related tasks. Figure 4 illustrates the transformer network architecture and Table 4 illustrates applications of transformer-based models.

Table 6.8: Applications of Transformer-Based Models

Application	Model
Machine Translation	Transformer, BERT (Bidirectional Encoder Representations from Transformers)
Question Answering	BERT, GPT (Generative Pre-trained Transformer)
Language Generation	GPT-2, GPT-3
Named Entity Recognition	BERT

Advances in deep learning and neural architectures have advanced the field of artificial intelligence, enabling breakthroughs in computer vision, natural language processing and other fields. This chapter explored major advances in CNN, RNN/LSTM, GAN, and transformer networks, presenting their applications and contributions to the state of the art. By leveraging these advances and improving existing architectures, researchers can continue to push the boundaries of deep learning and pave the way for smarter, more efficient systems.

6.2 Explainable AI and Interpretable Models

As AI systems become increasingly complex and pervasive, the need for transparency and interpretability increases. Explainable AI (XAI) aims to provide insight into the decision-making process of AI models, allowing users to understand and trust their outputs. This chapter explores the importance of explain ability in artificial intelligence and outlines various techniques and interpretable models that facilitate the understanding of artificial intelligence. In addition, related images and tables are provided to improve understanding and visualization of these concepts. 1. Importance of Artificial Intelligence Explained:

Explainable AI plays a defining role in areas where decision-making from AI models has a significant impact. For example, in healthcare, explainability is crucial to understanding the factors that influence a diagnosis or treatment recommendation. In addition, explainability increases accountability, fairness and trust in AI systems by allowing users to identify biases, errors or potential risks. Figure 1 illustrates the impact of explainability in different domains, while Table 1 illustrates the benefits of explainable AI.

Table 6.9: Benefits of Explainable AI

Benefit	Description
Transparency	Enables understanding of AI decision-making
Trust	Increases user confidence in AI systems

Accountability	Facilitates identification of biases and errors
Fairness	Helps detect and mitigate discriminatory outcomes
Compliance	Supports compliance with regulations and standards

A. Explainable artificial intelligence technologies:

Various techniques have been developed to explain the decisions made by AI models. These techniques range from post hoc methods that explain the behavior of the model after training to internally interpretable models. Prominent techniques include feature importance analysis, rule-based models, and model-agnostic approaches such as LIME (local interpretable model-agnostic explanations) and SHAP (Shapley adjunctive explanations). Figure 2 illustrates the use of LIME in clarifying image classification, while Table 2 provides an overview of different clarifying techniques.

Table 6.10: Overview of Explainability Techniques

Technique	Description
Feature Importance	Ranks input features based on their contribution to output
Rule-based Models	Represent decisions using interpretable rules
LIME (Local Interpretable Model-Agnostic Explanations)	Generates interpretable explanations for complex models
SHAP (Shapley Additive explanations)	Quantifies feature importance for a model's output

B. Models to be interpreted:

Interpretable models are designed with transparency and comprehensibility in mind and essentially provide interpretable representations or decision processes. Decision trees, linear models, and rule-based models are notable examples of interpretable models. Figure 3 shows the decision boundary of the decision tree and Table 3 shows different types of interpretable models and their applications.

Table 6.11: Types of Interpretable Models and Applications

Model	Applications
Decision Trees	Medical diagnosis, credit scoring
Linear Models	Risk assessment, fraud detection
Rule-based Models	Regulatory compliance, loan approval
Symbolic Models	Scientific discovery, expert systems

C. Evaluate Explainable AI: Evaluation criteria and frameworks must be established in order to evaluate the efficacy of explainable AI strategies. When assessing the justifications offered by AI models, proximity, fidelity, and comprehensibility are some of the important things to take into account. Evaluations of the usefulness and credibility of explanations also heavily rely on user studies and qualitative feedback. Table 4 lists crucial evaluation criteria and factors for explainable AI.

Table 6.12: Evaluation Metrics and Considerations for Explainable AI

Metric/Consideration	Description
Proximity	Measures the alignment between the explanation and model behavior
Fidelity	Assesses the faithfulness of the explanation to the model's decision-making process
Comprehensibility	Evaluates the clarity and understandability of the explanation
User Studies	Involves feedback from users to assess usefulness and trustworthiness

In order to ensure accountability, openness, and confidence in AI systems, explainable AI is essential. This chapter focused on the importance of explainability, covered several explainable AI strategies, and looked at interpretable models. Researchers and practitioners can address issues with biases, fairness, and ethical considerations by adding explainable AI practises into the creation and application of AI systems. The ethical and reliable implementation of AI technology will be facilitated by ongoing research and innovation in understandable AI.

6.3 Federated Learning and Privacy-Preserving ML:

With machine learning and data protection issues, new approaches are needed that balance the benefits of training collaborative models with the protection of sensitive data. Blended learning and privacy-preserving machine learning (ML) techniques address these challenges by enabling model training on distributed data while preserving individual privacy. This chapter introduces the concepts of blended learning and privacy-preserving ML, explores their applications, techniques, and related trade-offs. In addition, related images and tables are provided to improve understanding and visualization of these concepts.

A. Blended learning:

Federated learning makes it possible to train machine learning models on distributed devices or data sources without having to remove the raw data from their respective locations. The process involves repeated model updates where local devices or edge servers perform calculations based on their local data and share model updates only with the central server. Figure 1 illustrates the blended learning process, while Table 1 shows important blended learning applications.

Table 6.13: Applications of Federated Learning

Application	Description
Mobile Devices	Collaborative model training on smartphones
Healthcare	Privacy-preserving analysis of medical data

Internet of Things	Collaborative learning on edge devices
Financial Services	Fraud detection and risk assessment

B. Privacy-Preserving Machine Learning: This type of machine learning aims to safeguard private information while yet enabling model training or analysis. Common methods for preserving privacy include differential privacy, safe multiparty computation, and homomorphic encryption. The idea of differential privacy is demonstrated in Figure 2, and many privacy-preserving ML approaches are listed along with their uses in Table 2.

Table 6.14: Privacy-Preserving ML Techniques

Technique	Description
Homomorphic Encryption	Allows computation on encrypted data
Secure Multiparty Computation	Enables collaborative computation without revealing data
Differential Privacy	Quantifies and limits the privacy impact of data releases

C. Federated Learning Architectures: A number of distinct federated learning architectures have been put up to support various situations and data distributions. Federated distillation, federated reinforcement learning, and federated averaging are examples of these designs. Table 3 lists the various federated learning architectures and their features, whereas Figure 3 illustrates the federated averaging approach.

Table 6.15: Federated Learning Architectures

Architecture	Characteristics
Federated Averaging	Aggregates model updates using averaging
Federated Distillation	Transfers knowledge from a teacher model to local models
Federated Reinforcement Learning	Collaboratively learns policies through interactions

D. Trade-offs in federated learning and privacy-preserving ML:

Although blended learning and privacy-preserving ML offer valuable advantages, they also come with certain trade-offs. These trade-offs include increased communication costs, potential degradation of model performance due to limited data availability, and the complexity of ensuring privacy. Understanding these trade-offs helps make informed decisions when implementing blended learning and privacy-preserving ML approaches.

Thus Blended learning and privacy-preserving ML technologies offer the opportunity to collaborate on machine learning tasks while respecting data protection. This chapter explored the concepts of blended learning and privacy-preserving ML, as well as their applications, techniques, and tradeoffs. By implementing these approaches, organizations and researchers can harness the power of distributed data while complying with data protection regulations and individual rights. Continued advances in blended learning and privacy preserving ML will lead to an era of more secure and privacy machine learning.

6.4 Generative Models and Adversarial Learning

Generative models and competitive learning have revolutionized the field of artificial intelligence, enabling the creation of realistic and novel data. Applications for these models have been found in several fields such as image synthesis, text generation, and data augmentation. This chapter reviews the concepts of generative models and adversarial learning, emphasizing their architecture, training procedures, and applications. In addition, related images and tables are provided to improve understanding and visualization of these concepts.

A. Generative models:

Generative models aim to learn the underlying distribution of a given data set to generate new samples that resemble the original data. Notable generative models include adaptive autoencoders (VAEs) and generative adversarial networks (GANs). Figure 1 illustrates the architecture of VAE, while Table 1 illustrates various generation models and their applications.

Table 6.16: Generative Models and Applications

Model	Applications
Variational Autoencoders (VAEs)	Image synthesis, anomaly detection
Generative Adversarial Networks (GANs)	Image generation, style transfer
Autoregressive Models	Text generation, speech synthesis
Flow-based Models	Density estimation, data compression

B. Adversarial Learning: To learn and enhance the generative model, adversarial learning makes use of two competing networks, the discriminator and the generator. The discriminator seeks to discriminate between genuine and fraudulent samples, whereas the generator seeks to make examples that deceive it. Increasingly realistic training examples are produced by iterative training. The architecture of a GAN is illustrated in Figure 2, and various adversarial learning strategies and their applications are listed in Table 2.

Table 6.17: Adversarial Learning Techniques and Applications

Technique	Applications
Generative Adversarial Networks (GANs)	Image synthesis, style transfer
Adversarial Autoencoders	Image generation, data representation learning
Adversarial Domain Adaptation	Domain adaptation, unsupervised learning
Adversarial Reinforcement Learning	Reinforcement learning with model-free exploration

C. Generative Model Performance Evaluation: Analysing the effectiveness of generative models is a difficult issue. Inception Score, Fréchet Inception Distance (FID), and Perceptual Path Length (PPL) are examples of common measures. These metrics rank the generated samples' quality, diversity, and closeness to actual data. Important evaluation metrics and factors for generative models are listed in Table 3.

Table 6.18: Evaluation Metrics and Considerations for Generative Models

Metric/Consideration	Description
Inception Score	Measures quality and diversity of generated samples
Fréchet Inception Distance (FID)	Quantifies similarity between real and generated data distributions
Perceptual Path Length (PPL)	Evaluates the diversity of generated samples
User Studies	Involves feedback from users to assess the subjective quality of generated samples

D. Applications of Generative Models: Generative models have a wide range of uses in a number of different fields. These include style transfer, text production, data augmentation, and image synthesis.

While Table 4 lists the various uses of generative models, Figure 3 illustrates how they are used in picture synthesis.

Table 6.19: Applications of Generative Models

Application	Description
Image Synthesis	Generating realistic images
Data Augmentation	Expanding training datasets
Text Generation	Generating coherent and diverse text
Style Transfer	Transforming the style of an image or text

Hence by making it possible to generate new and realistic data, generative models and adversarial learning have revolutionized the area of artificial intelligence. The architectures, training processes, and applications of generative models and adversarial learning were highlighted in this chapter. Researchers and practitioners can push the limits of creativity, data augmentation, and content development in many different disciplines by utilizing the power of generative models.

6.5 Quantum Machine Learning

Quantum Machine Learning (QML) is an emerging field that combines the principles of quantum mechanics and machine learning to explore new computational paradigms. Quantum computers have the potential to outperform classical computers in certain computational tasks, leading to advances in areas such as optimization, pattern recognition and data analysis. This chapter introduces the concepts of quantum machine learning and highlights its fundamentals, algorithms, and applications. In addition, related images and tables are provided to improve understanding and visualization of these concepts.

A. Fundamentals of Quantum Machine Learning:

Understanding the principles of quantum mechanics is crucial to understanding quantum machine learning. Concepts such as superposition, entanglement, and quantum gates form the building blocks of quantum computing. Figure 1 illustrates the concept of superposition in quantum systems, while Table 1 outlines the key elements of quantum machine learning.

Table 6.20: Foundations of Quantum Machine Learning

Concept	Description
Superposition	Quantum systems can exist in multiple states simultaneously
Entanglement	Correlation between quantum systems
Quantum Gates	Operations that manipulate quantum states
Quantum Circuits	Sequences of quantum gates

B. Quantum Machine Learning Algorithms: Quantum Machine Learning algorithms use the strength of quantum computing to advance or develop existing machine learning strategies. Quantum Support Vector Machines (QSVM), Quantum Neural Networks (QNN), and Quantum Principal Component Analysis (QPCA) are notable algorithms. A quantum neural network's architecture is shown in Figure 2, and various quantum machine learning methods and their applications are presented in Table 2.

Table 6.21: Quantum Machine Learning Algorithms and Applications

Algorithm	Applications
Quantum Support Vector Machines (QSVM)	Classification, pattern recognition
Quantum Neural Networks	Function approximation, image recognition
Quantum Principal Component Analysis (QPCA)	Dimensionality reduction, data compression

C. Applications of Quantum Machine Learning: Quantum Machine Learning has the potential to have an impact in a variety of domains, including drug discovery, quantum chemistry, and optimization issues. Quantum Machine Learning's use in quantum chemistry is demonstrated in Figure 3, and its various applications are listed in Table 3.

Table 6.22: Applications of Quantum Machine Learning

Application	Description
Optimization Problems	Solving combinatorial optimization problems
Quantum Chemistry	Predicting molecular properties and behavior
Pattern Recognition	Image and speech recognition
Anomaly Detection	Identifying unusual patterns in data

D. Challenges and future directions:

Despite the promise of Quantum Machine Learning, there are several challenges that need to be addressed. These challenges include the need for error correction in quantum systems, limited qubit coherence time, and scarcity of quantum training data. Overcoming these challenges requires further development of quantum hardware, software and algorithmic techniques. Moreover, further research is needed to explore the potential of quantum machine learning in the fields of unsupervised learning and reinforcement learning.

Thus Quantum machine learning is an exciting field that combines the principles of quantum computing with machine learning techniques. This chapter explored the fundamentals, algorithms, and applications of quantum machine learning in various fields. Thanks to the continuous research and development of quantum hardware and algorithms, Quantum Machine Learning can revolutionize computing, enabling the solution of complex problems beyond classical computing.

6.6 Edge Computing and IoT Integration with Machine Learning

The proliferation of Internet of Things (IoT) devices and the demand for real-time computing has led to the emergence of edge computing. Edge computing brings computing power and intelligence closer to the data source, enabling faster response times, lower network latency and better privacy. This chapter explores the integration of edge computing and the Internet of Things with machine learning and highlights the benefits, challenges and applications. In addition, related images and tables are provided to improve understanding and visualization of these concepts.

A. Introduction to Edge Computing and IoT:

Edge computing involves deploying computing resources, including processing, storage and analysis, closer to the edge of the network where data is generated. IoT refers to a network of interconnected devices that communicate and share information with each other. Figure 1 illustrates the concept of edge computing and IoT integration, while Table 1 shows the main features of edge computing and IoT.

Table 6.23: Characteristics of Edge Computing and IoT

Characteristic	Description
Proximity to Data Source	Computing resources deployed closer to the data source, reducing latency
Real-Time Processing	Enables real-time data analysis and decision-making
Distributed Architecture	Devices and resources interconnected in a decentralized manner
Scalability and Flexibility	Easily scalable and adaptable to changing IoT environments
Privacy and Security	Data processing at the edge improves privacy and reduces data transmission risks

B. Machine Learning at the Edge: By combining machine learning with edge computing, it is possible to do data analysis and make intelligent decisions at edge devices. The reliance on cloud computing can be reduced by training machine learning models centrally and deploying them at the edge for local inference. Figure 2 shows how machine learning is used at the edge, and Table 2 lists machine learning methods that are appropriate for edge computing.

Table 6.24: Machine Learning Techniques for Edge Computing

Technique	Description
On-Device Machine Learning	Training and inference on edge devices
Transfer Learning	Fine-tuning pre-trained models at the edge
Federated Learning	Collaborative model training on edge devices
Lightweight Models	Small, efficient models suitable for edge devices

C. Advantages and Challenges of Edge Computing with ML:

The integration of edge computing and machine learning brings several benefits, such as lower latency, better response time, better privacy and security, and bandwidth savings. However, it also brings challenges such as limited computing resources, energy constraints, heterogeneous device characteristics and data heterogeneity. Understanding these benefits and challenges is essential to successfully deploying and managing machine learning-based edge computing.

D. Edge Computing and IoT Applications with ML:

The combination of edge computing, Internet of Things and machine learning opens up many applications in various fields. These applications include real-time monitoring and analysis, predictive maintenance, anomaly detection, intelligent monitoring and personalized healthcare. Figure 3 shows the application of edge computing and IoT integration in machine learning in healthcare, while Table 3 shows the various applications of this integration.

Table 6.25: Applications of Edge Computing and IoT Integration with ML

Application	Description
Predictive Maintenance	Detecting and preventing equipment failures
Intelligent Transportation	Real-time traffic analysis and optimization
Smart Grids	Real-time energy management and optimization
Environmental Monitoring	Real-time analysis of air and water quality
Smart Healthcare	Remote patient monitoring and diagnosis

Integrating edge computing and the Internet of Things with machine learning offers transformative opportunities for real-time data processing, intelligent decision-making and greater efficiency. This chapter explored the concepts of integrating edge computing and IoT with machine learning and highlighted their benefits, challenges and applications. By leveraging extreme computing and machine learning, organizations can unleash the full potential of IoT devices, enabling innovative solutions across industries.

6.7 Ethics and Fairness

Machine learning algorithms have enormous potential to revolutionize various domains. However, as these algorithms become more common, ethical considerations must be taken into account and the fairness of their implementation must be ensured. This chapter explores the ethical implications of machine learning and the importance of fairness in algorithmic decision making. It introduces key ethical principles, measures of fairness and approaches to mitigate bias. In addition, related images and tables are provided to improve understanding and visualization of these concepts. 1. Ethical aspects of machine learning:

Because machine learning algorithms make decisions that affect individuals and society, it is important to consider ethical considerations. The main ethical principles are transparency, accountability, privacy and fairness. Figure 1 illustrates the ethical aspects of machine learning and Table 1 illustrates the principles and their implications.

Table 6.26: Ethical Principles in Machine Learning

Ethical Principle	Implications
Transparency	Ensuring algorithms and decision-making processes are explainable and understandable
Accountability	Holding individuals and organizations responsible for the outcomes of machine learning systems
Privacy	Safeguarding the personal information and data of individuals
Fairness	Ensuring unbiased and equitable treatment across different groups and demographics

A. Fairness in Machine Learning: Results that are biassed or discriminatory may have significant societal and economic repercussions. Fairness is a critical component of machine learning. The fairness of algorithms is evaluated using fairness measures like differential impact, equal opportunity, and predictive parity. The idea of fairness in machine learning is illustrated in Figure 2, and fairness measurements and their implications are presented in Table 2.

Table 6.27: Fairness Metrics in Machine Learning

Fairness Metric	Description
Disparate Impact	Measures the difference in outcomes across different groups
Equal Opportunity	Measures the true positive rate across different groups

Predictive Parity	Measures the positive predictive value across different groups

B. Addressing Bias and Reducing Inequity: Machine learning methods are prone to biases that might reinforce unfair results. To maintain fairness, it is critical to recognise and deal with these prejudices. Pre-processing, in-processing, and post-processing procedures are all used to reduce bias. While Table 3 lists bias reduction strategies and their effects, Figure 3 illustrates the machine learning approach for bias mitigation.

Table 6.28: Bias Mitigation Techniques in Machine Learning

Bias Mitigation Technique	Description
Pre-processing	Modifying the training data to reduce biases
In-processing	Modifying the learning process to reduce biases
Post-processing	Modifying the predictions to reduce biases

C. Applications and Case Studies: Fairness and ethics in machine learning have an impact across many fields. The importance of addressing ethical issues and advancing fairness is highlighted via case studies and applications from real-world situations. Table 4 lists various areas and instances where machine learning ethics and fairness are important.

Table 6.29: Applications and Case Studies

Domain	Examples
Hiring and Recruitment	Reducing bias in resume screening and candidate selection processes

Criminal Justice	Ensuring fairness in risk assessment and sentencing algorithms
Healthcare	Avoiding biases in diagnostic and treatment recommendation systems
Financial Services	Fairness in credit scoring and loan approval processes

The ethical ramifications of machine learning must be addressed in order to maintain fairness in algorithmic decision-making as it gets more and more ingrained in our daily lives. This chapter looked at how to eliminate biases, the value of fairness, and ethical issues in machine learning. Organizations can encourage the responsible and fair use of machine learning algorithms across a range of disciplines by including ethical standards and fairness measures.

6.8 Human-Centric Machine Learning

As machine learning technologies continue to evolve, it is critical to focus on the human side of these systems. Human-centered machine learning emphasizes collaboration between humans and machines, taking into account human values, preferences, and interpretability. This chapter examines the principles, methods, and applications of human-centered machine learning. It explores the importance of interpretability, user-centered design, and human feedback. In addition, related images and tables are provided to improve understanding and visualization of these concepts. 1. Principles of human-centered machine learning:

Human-centered machine learning aims to adapt machine learning algorithms to the needs and values of people. This approach is based on the principles of transparency, interpretability and explainability. Figure 1 illustrates the principles of human-centered machine learning, and Table 1 lists the most important principles and their impact.

Table 6.30: Principles of Human-Centric Machine Learning

Principle	Implications
Transparency	Providing insights into the decision-making process
Interpretability	Understanding how and why the model makes predictions
Explainability	Providing explanations for model predictions

A. User-Centric Design: Machine learning systems must be designed with the end user in mind if they are to be accepted and used. Usability, accessibility, and user experience are the three main areas of user-centric design. Table 2 lists important factors and approaches for user-centric design, whereas Figure 2 illustrates the user-centric design process in Human-Centric Machine Learning.

Table 6.31: Considerations and Methodologies for User-Centric Design

Consideration/Methodology	Description
User Research	Gathering insights into user needs and preferences
Prototyping	Iterative development of user interfaces and interactions
Usability Testing	Evaluating the system's ease of use and user satisfaction
Accessibility	Ensuring the system is usable by individuals with disabilities
Human-Computer Interaction (HCI)	Incorporating HCI principles for effective human interaction

B. Interpretability and Explainability: In order to foster trust and understanding between people and machine learning models, interpretability and explainability are crucial. To achieve interpretability and explainability, a variety of strategies can be used, such as rule extraction, attention mechanisms, and model-agnostic techniques. Human-Centric Machine Learning's interpretability and explainability methodologies are demonstrated in Figure 3 and their implications are listed in Table 3.

Table 6.32: Interpretability and Explainability Methods

Method	Description
Model-Agnostic Techniques	Post-hoc methods that explain model behavior independently of the learning algorithm
Rule Extraction	Extracting interpretable rules from complex models
Attention Mechanisms	Identifying relevant features or components for decision-making
Counterfactual Explanations	Providing alternative scenarios to understand the model's behavior

C. Human Feedback and Iterative Improvement: Integrating human feedback into machine learning systems is essential for improving them and bringing them closer to the needs and values of people. Key elements of this technique include feedback loops, user-driven customization, and iterative improvement. The methods and approaches for incorporating human feedback into human-centric machine learning are shown in Table 4.

Table 6.33: Approaches for Incorporating Human Feedback

Approach	Description
Feedback Loops	Collecting feedback from users to refine models
Active Learning	Iteratively selecting informative data for labeling
Collaborative Filtering	Incorporating user preferences and recommendations

Customization and Personalization	Allowing users to customize models and preferences

D. Applications of Human-Centric Machine Learning: Human-Centric Machine Learning has a wide range of uses, from finance and entertainment to healthcare and education. Case studies from real-world situations show how adopting human-centric strategies is advantageous. Table 5 lists various fields and instances where human-centric machine learning is vital.

Table 6.34: Applications of Human-Centric Machine Learning

Domain	Examples
Healthcare	Personalized treatment recommendations based on patient preferences
Education	Adaptive learning systems tailored to individual student needs
Finance	Explainable credit scoring models for transparency and trust
Entertainment	Recommendation systems that consider user preferences and interests

Interpretability, user-centric design, and human feedback are prioritized in human-centered machine learning, which emphasizes the interaction between humans and machines. The foundational ideas, approaches, and practical uses of human-centric machine learning were examined in this chapter. Organizations may design machine learning systems that are more visible, interpretable, and in line with people's demands by taking into account their values and preferences.

6.9 Reinforcement Learning in Real-World Applications

Reinforcement learning (RL) is a powerful approach to machine learning that allows agents to learn optimal actions while interacting with the environment. This chapter explores the application of RL in real-world scenarios, highlighting its potential, challenges, and advances. It introduces various RL algorithms, discusses their applicability in various domains, and provides an overview of the practical aspects of implementing RL systems. In addition, related images and tables are provided to improve understanding and visualization of these concepts.

A. Introduction to Reinforcement Learning:

Reinforcement learning is a branch of machine learning that focuses on training agents to make sequential decisions in an environment to maximize cumulative reward. RL agents learn through trial and error, using feedback from the environment to improve their decision-making practices. Figure 1 provides an overview of the RL framework, while Table 1 provides the key concepts and components of RL.

Table 6.35: Key Concepts and Components of RL

Concept/Component	Description
Agent	Learns to interact with the environment and take actions
Environment	Provides the state, feedback, and rewards to the agent
State	Represents the current state of the environment
Action	Decisions made by the agent to transition between states
Reward	Feedback signal that indicates the desirability of an action

B. RL Techniques & Algorithms:

To address various difficulties, many RL algorithms and strategies have been created. An overview of popular RL algorithms, such as Q-

Learning, Policy Gradient techniques, and Proximal Policy Optimisation (PPO), is given in this section. Different RL algorithms are shown in Figure 2, and their features and application domains are listed in Table 2.

Table 6.36: RL Algorithms and their Application Domains

Algorithm	Characteristics	Application Domains
Q-Learning	Value-based, learns Q-values	Robotics, Game playing, Control systems
Policy Gradient Methods	Policy-based, learns policy directly	Robotics, Natural Language Processing, Game playing
Proximal Policy Optimization (PPO)	Policy-based, improves stability and convergence	Robotics, Autonomous Vehicles, Simulations

C. Difficulties with RL Applications in the Real World:

Numerous difficulties arise when RL algorithms are used in practical settings. These include the ineffectiveness of samples, the trade-off between exploration and exploitation, reward engineering, and safety issues. For deployment to be effective, it is essential to recognise and overcome these obstacles. Table 3 lists the difficulties and factors to take into account in real-world RL applications.

Table 6.37: Challenges and Considerations in Real-World RL Applications

Challenge	Description
Sample Inefficiency	High sample complexity and data requirements
Exploration-Exploitation Trade-off	Balancing between exploring new actions and exploiting known ones

Reward Engineering	Designing appropriate reward functions
Safety and Ethics	Ensuring RL systems operate within defined safety boundaries and ethical considerations

D. RL in the Real World: RL has found use in a variety of fields, from robotics and autonomous systems to finance and healthcare. Case examples from the real world demonstrate the effectiveness of RL in resolving complicated issues. Table 4 lists various domains and illustrations of RL applications.

Table 6.38: Real-World Applications of RL

Domain	Examples
Robotics	Autonomous navigation, grasping and manipulation
Autonomous Vehicles	Self-driving cars, traffic optimization
Finance	Portfolio management, algorithmic trading
Healthcare	Personalized treatment recommendations

Real-world applications for reinforcement learning are excitingly possible since it allows agents to develop optimal behaviors through interactions with their surroundings. This chapter examined the foundations of RL, emphasized well-liked algorithms and strategies, and covered the difficulties and factors to be taken into account in practical RL applications. Organizations can use RL to tackle challenging issues and make outstanding progress in a variety of fields.

6.10 The Future of Machine Learning

Machine learning (ML) has made significant advances in recent years and has revolutionized several industries and fields. This chapter examines the future of ML, discusses emerging trends, challenges, and possible directions for future development. It explores cutting-edge technologies such as deep learning, generative models, and reinforcement learning, and outlines their future applications and implications. In addition, related images and tables are provided to improve understanding and visualization of these concepts.

A. Emerging Trends in Machine Learning:

The field of ML continues to evolve rapidly, with many emerging trends shaping its future. This section highlights key trends including:

B. Deep Learning and Neural Architectures:

Deep learning has taken ML to new heights, enabling the development of complex models that excel at tasks such as image recognition, natural language processing and speech synthesis. The future of ML will witness advances in neural architecture, including more efficient models, learning techniques, and integration with other fields.

C. Explainable artificial intelligence and interpretable models:

As ML algorithms become increasingly complex, the need for interpretability and explainability increases. In the future, the focus will be on creating interpretable ML models that can provide transparent explanations for their decision-making processes. It facilitates trust, accountability and ethical considerations in ML systems.

D. Federated Learning and Privacy Preserving ML:

Privacy concerns are at the forefront of ML development. Blended learning and privacy-preserving techniques are highlighted that enable models to be trained on distributed devices without compromising user data. This approach enables collaboration and information exchange while ensuring data protection.

E. Generative models and adversarial learning:

Generative models such as Generative Adversarial Networks (GAN) have revolutionized image synthesis and data generation. Future advances will focus on improving creative models, improving their stability, and extending applications to areas such as video synthesis, text generation, and virtual reality.

F. Quantum Machine Learning:

Quantum computing has enormous potential for ML, providing exponential computing power and the ability to solve complex optimization problems. Quantum machine learning algorithms are emerging, enabling advances in areas such as drug discovery, optimization and cryptography.

G. Integrating Edge Computing and IoT with ML:

The proliferation of Internet of Things (IoT) devices and the need for real-time decision-making is driving the integration of ML in edge computing. Future focus will be on deploying ML models directly at the edge, reducing latency, improving privacy and enabling efficient data processing.

H. Challenges and ethical aspects:

Although the future of ML is promising, it also brings with it a number of challenges and ethical considerations. They include:

1. Illusion and Justice:

A critical challenge is to address biases in ML algorithms and ensure fairness in populations and groups. In the future, we will focus on reducing biases, developing fairer algorithms and promoting ethical aspects in ML systems.

2. Data quality and data protection:

Data quality and privacy remain vital concerns. Ensuring high-quality, diverse and representative data for training ML models while protecting user privacy is a top priority. Future developments will explore methods to anonymize data, create synthetic data, and maintain data privacy.

3. Interpretability and explain-ability:

Interpreting and explaining decisions made with ML models is critical for trust and accountability. Future research will focus on developing more interpretable models, improving explanatory techniques, and providing a transparent view of the decision-making process.

4. Additional instructions and applications:

The future of ML has enormous potential for change in many different areas. This section discusses possible guidelines and applications, including:

I. Health care:

ML plays a crucial role in personalized medicine, disease diagnosis, drug development and precision medicine. Advances in medical imaging, genomics and clinical decision support systems are revolutionizing healthcare practices.

1. Autonomous systems:

In the development of autonomous vehicles, robotics and drones, ML is used for observation, decision making and control. Future advances will focus on improving the safety, efficiency and adaptability of autonomous systems.

2. Natural language processing:

Advances in natural language processing enable improved language understanding, emotional analysis and language creation. Further developments will enable seamless human-computer interaction, language translation and content creation.

3. Climate change and sustainability:

ML plays a crucial role in combating climate change, optimizing energy consumption and improving sustainability. Future applications include climate modeling, renewable energy optimization, and resource management.

The future of ML offers significant opportunities driven by emerging trends, technological advances and the need for ethical considerations. This chapter explored potential trends and applications of ML, discussed emerging trends, challenges and ethical considerations. By staying abreast of these developments, organizations can harness the transformative power of ML and shape the future of various industries and sectors.

9789348037879